LAB MANUAL

METALLURGY FUNDAMENTALS

Ferrous and Nonferrous

Seventh Edition

J.C. Warner • R. Dean Odell

Publisher
The Goodheart-Willcox Company, Inc.
Tinley Park, IL
www.g-w.com

Copyright © 2025
by
The Goodheart-Willcox Company, Inc.

All rights reserved. No part of this work may be reproduced, stored, or transmitted in any form or by any electronic or mechanical means, including information storage and retrieval systems, without the prior written permission of
The Goodheart-Willcox Company, Inc.

ISBN 979-8-89118-155-7

1 2 3 4 5 6 7 8 9 – 25 – 28 27 26 25 24 23

The Goodheart-Willcox Company, Inc. Brand Disclaimer: Brand names, company names, and illustrations for products and services included in this text are provided for educational purposes only and do not represent or imply endorsement or recommendation by the authors or the publisher.

The Goodheart-Willcox Company, Inc. Safety Notice: The reader is expressly advised to carefully read, understand, and apply all safety precautions and warnings described in this book or that might also be indicated in undertaking the activities and exercises described herein to minimize risk of personal injury or injury to others. Common sense and good judgment should also be exercised and applied to help avoid all potential hazards. The reader should always refer to the appropriate manufacturer's technical information, directions, and recommendations; then proceed with care to follow specific equipment operating instructions. The reader should understand these notices and cautions are not exhaustive.

The publisher makes no warranty or representation whatsoever, either expressed or implied, including but not limited to equipment, procedures, and applications described or referred to herein, their quality, performance, merchantability, or fitness for a particular purpose. The publisher assumes no responsibility for any changes, errors, or omissions in this book. The publisher specifically disclaims any liability whatsoever, including any direct, indirect, incidental, consequential, special, or exemplary damages resulting, in whole or in part, from the reader's use or reliance upon the information, instructions, procedures, warnings, cautions, applications, or other matter contained in this book. The publisher assumes no responsibility for the activities of the reader.

The Goodheart-Willcox Company, Inc. Internet Disclaimer: The Internet resources and listings in this Goodheart-Willcox Publisher product are provided solely as a convenience to you. These resources and listings were reviewed at the time of publication to provide you with accurate, safe, and appropriate information. Goodheart-Willcox Publisher has no control over the referenced websites and, due to the dynamic nature of the Internet, is not responsible or liable for the content, products, or performance of links to other websites or resources. Goodheart-Willcox Publisher makes no representation, either expressed or implied, regarding the content of these websites, and such references do not constitute an endorsement or recommendation of the information or content presented. It is your responsibility to take all protective measures to guard against inappropriate content, viruses, or other destructive elements.

Image Credits. Front cover, left: Pavel Nesvadba/Shutterstock.com; top right: Andriy Solovyov/Shutterstock.com; middle right: Matveev Aleksandr/Shutterstock.com; bottom right: Monstar Studio/Shutterstock.com

Introduction

This lab manual is designed for use with the text *Metallurgy Fundamentals*. The labs expand on ideas from several chapters that apply to practical, real-world testing and experiences relating to metallurgy and metallurgical fields.

Each lab contains hands-on lab activities and review questions that enhance textbook content and allow you to observe, test, experience, and reflect upon metallurgical properties and practices firsthand. Two examples of hardness charts are included at the beginning of the lab manual for convenient reference.

Reading *Metallurgy Fundamentals* and using this lab manual will help you acquire a working knowledge of the principles and processes of metallurgy and their application. Answering the questions for each lab will help you understand and master the technical knowledge presented in the text. The activities range from chapter content reinforcement to real-world application, including activities related to different metallurgical disciplines. It is important in these activities to understand any safety procedures set forth by your instructor.

Contents

Hardness Conversion Charts v

Lab 1
Hardness Testing of Steel 1

Lab 2
Measuring Rockwell Hardness with Different Scales 7

Lab 3
Determining Hardness and Alloy Type by Spark and File Tests. 11

Lab 4
Hardness Testing of Thin Sheet 17

Lab 5
Hardness Testing of Castings 23

Lab 6
Tensile Testing of Steel 31

Lab 7
Metal or Nonmetal? 37

Lab 8
Metallurgical Specimen Preparation 47

Lab 9
Ductile-to-Brittle Transition Temperature of Steel 51

Lab 10
Hardenability of Steel 61

Lab 11
Tempering Martensite 65

Lab 12
Effects of Annealing on Cold-Worked Brass 69

Lab 13
Determining Strength of a Weld 75

Lab 14
Age Hardening of Aluminum 79

Lab 15
Pack Carburizing of Steel 87

Lab 16
Strain-Hardening Copper Wire 91

Hardness Conversion Charts

Typical Rockwell B and Rockwell C hardness conversion charts for non-austenitic steels are provided here for convenient use. Remember that with such charts, the relation of tensile strength to hardness is inexact.

Hardness Conversion Chart for Non-Austenitic Steels (Rockwell B Hardness Range)

Rockwell B	Rockwell Superficial 15-T	Rockwell Superficial 30-T	Rockwell Superficial 45-T	Brinell 3000 kg	Tensile Strength ksi*
100	93.1	83.1	72.9	240	116
99	92.8	82.5	71.9	234	114
98	92.5	81.8	70.9	228	109
97	92.1	81.1	69.9	222	104
96	91.8	80.4	68.9	216	102
95	91.5	79.8	67.9	210	100
94	91.2	79.1	66.9	205	98
93	90.8	78.4	65.9	200	94
92	90.5	77.8	64.8	195	92
91	90.2	77.1	63.8	190	90
90	89.9	76.4	62.8	185	89
89	89.5	75.8	61.8	180	88
88	89.2	75.1	60.8	176	86
87	88.9	74.4	59.8	172	84
86	88.6	73.8	58.8	169	83
85	88.2	73.1	57.8	165	82
84	87.9	72.4	56.8	162	81
83	87.6	71.8	55.8	159	80
82	87.3	71.1	54.8	156	77
81	86.9	70.4	53.8	153	73
80	86.6	69.7	52.8	150	72
79	86.3	69.1	51.8	147	70
78	86.0	68.4	50.8	144	69
77	85.6	67.7	49.8	141	68
76	85.3	67.1	48.8	139	67
75	85.0	66.4	47.8	137	66
74	84.7	65.7	46.8	135	65

*Note: The tensile strength relation to hardness is inexact, even for steel, unless it is determined for a specific alloy.

Goodheart-Willcox Publisher

Continued

Hardness Conversion Chart for Non-Austenitic Steels (Rockwell B Hardness Range) Continued

Rockwell B	Rockwell Superficial 15-T	Rockwell Superficial 30-T	Rockwell Superficial 45-T	Brinell 3000 kg	Tensile Strength ksi*
73	84.3	65.1	45.8	132	64
72	84.0	64.4	44.8	130	63
71	83.7	63.7	43.8	127	62
70	83.4	63.1	42.8	125	61
69	83.0	62.4	41.8	123	60
68	82.7	61.7	40.8	121	59
67	82.4	61.0	39.8	119	58
66	82.1	60.4	38.7	117	57
65	81.8	59.7	37.7	116	56
64	81.4	59.0	36.7	114	—
63	81.1	58.4	35.7	112	—
62	80.8	57.7	34.7	110	—
61	80.5	57.0	33.7	108	—
60	80.1	56.4	32.7	107	—
59	79.8	55.7	31.7	106	—
58	79.5	55.0	30.7	104	—
57	79.2	54.4	29.7	103	—
56	78.8	53.7	28.7	101	—
55	78.5	53.0	27.7	100	—
54	78.2	52.4	26.7	—	—
53	77.9	51.7	25.7	—	—
52	77.5	51.0	24.7	—	—
51	77.2	50.3	23.7	—	—
50	76.9	49.7	22.7	—	—
49	76.6	49.0	21.7	—	—
48	76.2	48.3	20.7	—	—
47	75.9	47.7	19.7	—	—
46	75.6	47.0	18.7	—	—
45	75.3	46.3	17.7	—	—
44	74.9	45.7	16.7	—	—
43	74.6	45.0	15.7	—	—
42	74.3	44.3	14.7	—	—
41	74.0	43.7	13.6	—	—
40	73.6	43.0	12.6	—	—

*Note: The tensile strength relation to hardness is inexact, even for steel, unless it is determined for a specific alloy.

Continued

Goodheart-Willcox Publisher

Hardness Conversion Chart for Non-Austenitic Steels (Rockwell B Hardness Range) Continued

Rockwell B	Rockwell Superficial 15-T	Rockwell Superficial 30-T	Rockwell Superficial 45-T	Brinell 3000 kg	Tensile Strength ksi*
39	73.3	42.3	11.6	—	—
38	73.0	41.6	10.6	—	—
37	72.7	41.0	9.6	—	—
36	72.3	40.3	8.6	—	—
35	72.0	39.6	7.6	—	—
34	71.7	39.0	6.6	—	—
33	71.4	38.3	5.6	—	—
32	71.0	37.6	4.6	—	—
31	70.7	37.0	3.6	—	—
30	70.4	36.3	2.6	—	—

*Note: The tensile strength relation to hardness is inexact, even for steel, unless it is determined for a specific alloy.

Goodheart-Willcox Publisher

Hardness Conversion Chart for Non-Austenitic Steels (Rockwell C Hardness Range)

Rockwell C	Rockwell Superficial 15-T	Rockwell Superficial 30-T	Rockwell Superficial 45-T	Brinell 3000 kg	Tensile Strength ksi*
65	—	—	—	739	—
64	—	—	—	722	—
63	—	—	—	706	—
62	—	—	—	688	—
61	—	—	—	670	—
60	—	—	—	654	—
59	—	—	—	634	351
58	—	—	—	615	338
57	—	—	—	595	325
56	—	—	—	577	313
55	—	—	—	560	301
54	—	—	—	543	292
53	—	—	—	525	283
52	—	—	—	512	273
51	—	—	—	496	264
50	—	—	—	481	255
49	—	—	—	469	246
48	—	—	—	455	238
47	—	—	—	443	229

*Note: The tensile strength relation to hardness is inexact, even for steel, unless it is determined for a specific alloy.
Continued

Goodheart-Willcox Publisher

Hardness Conversion Chart for Non-Austenitic Steels (Rockwell C Hardness Range) Continued

Rockwell C	Rockwell Superficial 15-T	Rockwell Superficial 30-T	Rockwell Superficial 45-T	Brinell 3000 kg	Tensile Strength ksi*
46	—	—	—	432	221
45	—	—	—	421	215
44	—	—	—	409	208
43	—	—	—	400	201
42	—	—	—	390	194
41	—	—	—	381	188
40	—	—	—	371	182
39	—	—	—	362	177
38	—	—	—	353	171
37	—	—	—	344	166
36	—	—	—	336	161
35	—	—	—	327	156
34	—	—	—	319	152
33	—	—	—	311	149
32	—	—	—	301	146
31	—	—	—	294	141
30	—	—	—	286	138
29	—	—	—	279	135
28	—	—	—	271	131
27	—	—	—	264	128
26	—	—	—	258	125
25	—	—	—	253	123
24	—	—	—	247	119
23	—	—	—	243	117
22	—	—	—	237	115
21	—	—	—	231	112
20	—	—	—	226	110

*Note: The tensile strength relation to hardness is inexact, even for steel, unless it is determined for a specific alloy.

Goodheart-Willcox Publisher

Name _____ Date _____ Class _____

Hardness Testing of Steel

Introduction

Hardness testing is a good way to estimate tensile strength quickly and easily. The procedure for hardness testing is discussed in Chapter 4 of the *Metallurgy Fundamentals* textbook.

Objectives

- Learn to make hardness measurements with a Rockwell hardness test machine using the Rockwell HRB scale.
- Using a Rockwell hardness tester and scale, determine any difference in hardness between three steel alloys.

Safety Considerations

- Keep hands away from the test area while the hardness tester is in operation.
- Wear safety glasses when tests are being conducted.
- Ensure safe handling of any heavy specimens.

Equipment

- A Rockwell tester with dial gauge readout, **Figure 1-1**. Newer units will have a digital display. Your instructor will set up the tester for the Rockwell B scale. This requires a weight to apply 100 kg force on a 1/16″ diameter ball tungsten carbide indenter.

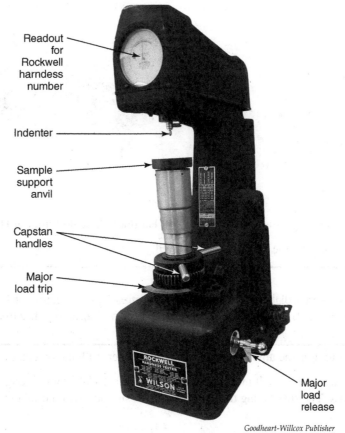

Goodheart-Willcox Publisher
Figure 1-1. A Rockwell tester with a dial gauge readout.

- Smartphone calculator (or any calculator)

Materials

- One sample each of three steel alloys of hot-rolled bar stock: AISI 1020, AISI 1045, and AISI 1095. AISI 1020 steel is iron with 0.2% carbon, 1045 contains 0.45% carbon, and 1095 steel contains 0.95% carbon.

Procedure

1. First, check the samples for suitability. The flat ends should be parallel and more or less square. The ends should be polished or ground smooth—at least 600 grit. There may be small indents from previous labs, **Figure 1-2**. The alloy of each sample must be clearly identified.

Goodheart-Willcox Publisher

Figure 1-2. Check the samples for suitability.

2. Label each sample, marking the side with a paint marker or engraver. *Do not* label the area that will be indented, **Figure 1-3**.

Goodheart-Willcox Publisher

Figure 1-3. Label each sample.

3. Check that the indenter and weights are set for Rockwell Hardness B scale (units of HRB).

4. Use the flat anvil for this lab, *not* the V-shaped (Vee) anvil.

5. **NOTE**
   This instruction is for a manual, dial gauge-type Rockwell tester. Your instructor will have instructions if your machine is a different model. The automatic operations shown in Chapter 4 of the textbook are followed by all Rockwell testers.

   Turn the capstan handles to drop the anvil until the sample fits underneath the indenter easily.

6. Position the sample so the indenter will indent a flat surface at least 1/8″ from all edges and 1/8″ from any previous hardness indents. The side facing the anvil should be smooth, with no indents. The sample must not rock on the anvil.

7. Raise the sample by turning the capstan, stopping just before the sample touches the indenter. Slowly turn further until the smaller dial matches up with the small dot, **Figure 1-4**. This applies the *minor load*. Rotate the outer dial until the large dial is lined up with the zero set-point.

Name _____

8. Trip the release to allow the tester to slowly apply the *major load*. When the dial settles on a constant reading—about 30 seconds—pull the major force release lever toward you to release the major load. Move slowly in both operations; do not jerk the machine.

9. Read the hardness value for this sample on the dial face for the scale you are using, and record it below. Hand off the sample to another student, or use another sample of the same alloy. You will need at least three hardness measurements for each alloy, and every student should make multiple hardness readings. More than three readings per alloy is fine.

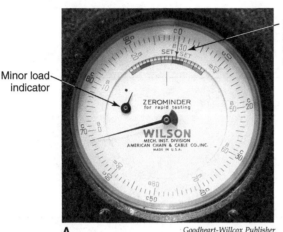

Figure 1-4. A—Hardness tester dial face, showing small dial for applying the minor load and adjustable marked ring for major load setting. B—The large dial is lined up with the zero set-point, ready for testing.

First Hardness Reading
Sample Alloy: AISI 1020

Test Number: _____

Rockwell B Hardness Reading: _____

Notes: _____

Second Hardness Reading
Sample Alloy: AISI 1020

Test Number: _____

Rockwell B Hardness Reading: _____

Notes: _____

Third Hardness Reading
Sample Alloy: AISI 1020

Test Number: _____

Rockwell B Hardness Reading: _____

Notes: _____

Extra Hardness Reading
Sample Alloy: AISI 1020

Test Number: _____

Rockwell B Hardness Reading: _____

Notes: _____

10. Take the same number of readings with the second and third alloys and record them below. On the Rockwell B scale, the reading for the 1095 sample may slightly exceed 100. This is OK for this lab because you need to use the same scale to compare the results across all samples.

First Hardness Reading
Sample Alloy: AISI 1045

Test Number: _____

Rockwell B Hardness Reading: _____

Notes: _____

Second Hardness Reading
Sample Alloy: AISI 1045

Test Number: _____

Rockwell B Hardness Reading: _____

Notes: _____

Third Hardness Reading
Sample Alloy: AISI 1045

Test Number: _____

Rockwell B Hardness Reading: _____

Notes: _____

Extra Hardness Reading
Sample Alloy: AISI 1045

Test Number: _____

Rockwell B Hardness Reading: _____

Notes: _____

First Hardness Reading
Sample Alloy: AISI 1095

Test Number: _____

Rockwell B Hardness Reading: _____

Notes: _____

Second Hardness Reading
Sample Alloy: AISI 1095

Test Number: _____

Rockwell B Hardness Reading: _____

Notes: _____

Name _____

Third Hardness Reading

Sample Alloy: AISI 1095

Test Number: _____

Rockwell B Hardness Reading: _____

Notes: _____

Extra Hardness Reading

Sample Alloy: AISI 1095

Test Number: _____

Rockwell B Hardness Reading: _____

Notes: _____

11. Continue making hardness readings until you have at least three readings for each alloy.

Review Questions

1. When you have completed measurements on all samples, look at your results. Are the measurements of the 1020, 1045, and 1095 samples generally different?

   _____

   _____

   _____

2. Use the calculator on your smartphone to find the average hardness of each alloy, and record your results below.

$$\text{Average hardness} = \frac{\text{Measurement 1} + \text{Measurement 2} + ... + \text{Measurement last}}{\text{Number of measurements}}$$

$$\text{HRB}_{Average} = \frac{\text{HRB}_1 + \text{HRB}_2 + ... + \text{HRB}_n}{n}$$

Example:

$$\frac{95.5 + 96.5 + 94.5}{3} = 95.5$$

Average Hardness of Samples

Alloy AISI 1020 (0.20% carbon): _____

Alloy AISI 1045 (0.45% carbon): _____

Alloy AISI 1095 (0.95% carbon): _____

3. Are the average hardness values of the 1020, 1045, and 1095 alloys much different from each other or nearly the same?

   _____

   _____

   _____

4. These samples were processed the same way, so they are only different in carbon composition. Can you see a trend of hardness with composition? What is that trend, if you find one?

   _____
   _____
   _____
   _____

5. Typical hardness readings by experienced people are within ±1.5 points of the average most of the time. Were your readings within that range? If you have any readings more than 4 hardness points away from the alloy average, discuss how they were made with your instructor. Can you identify possible reasons for these "extreme" values?

   _____
   _____
   _____
   _____
   _____
   _____

Name _____ Date _____ Class _____

Measuring Rockwell Hardness with Different Scales

Introduction
Steel can change its properties depending on processing (see Chapters 10 through 13 of the textbook). We need a way to measure such changes. Various settings of the Rockwell hardness tester are quick and fairly simple to use to get a measure of hardness. From *Lab 1: Hardness Testing of Steel*, you know that hardness relates to tensile strength.

Objectives
- Learn to use a Rockwell tester at different test configurations.
- Compare steel heat treated by different quenchants.

Safety Considerations
- Be properly respectful of hot furnaces and hot samples.
- Always be aware of how you would exit a furnace area if something should get out of whack.
- Use tongs and heavy gloves to handle hot samples(!).
- Wear safety glasses.

Equipment
- Furnace for heat treatment
- Heavy gloves and tongs to handle hot samples
- Rockwell hardness tester
- Three small buckets to use for the three cooling systems: water, oil, and air
- Two Rockwell test indenters: the diamond braille and the 1/16″ ball
- A black permanent marking pen

Materials
- At least four plain carbon steel specimens of a single heat-treatable carbon steel alloy, 3/4″ diameter and 3/4″ long
- 600 grit polishing paper, mounted so the flat sides of the samples can be cleaned

Procedure

1. Place three samples in a preheated furnace at 1600°F (871°C) for 30 minutes. When they are fully hot, the samples are the same color as the furnace walls. Pull each sample out of the furnace and quench one by water, one by oil, and one by air cooling. The fourth specimen will be a control.
2. Set up the Rockwall tester for the B scale. The settings are given in the Rockwell Hardness Scales table.

Rockwell Hardness Scales				
(From ASTM E18-22, Standard Test Methods for Rockwell Hardness of Metallic Materials)				
Scale Symbol	Indenter	Total Test Force, kgf	Dial Figures	Typical Applications of Scales
B	1/16" (1.588 mm) ball diamond	100	red	Copper alloys, **soft steels**, aluminum alloys, malleable iron, etc.
C	diamond	150	black	**Steel**, hard cast irons, pearlitic malleable iron, titanium, deep case-hardened steel, and other **materials harder than B100**.

Goodheart-Willcox Publisher

Figure 2-1. Use this chart to set up your Rockwell tester for the B scale.

3. As you take each cooled sample out of the quenchant, mark on a round side W, O, or A, depending on the quenchant, so you can tell them apart later. Record the applicable information as you go along.
4. Clean off the flat sides of each sample, removing any scale, and smooth them with the 600 grit paper. Make hardness measurements only on one side of each sample. Remember to keep each hardness indent about 1/8" away from the others.
5. Make three hardness measurements on each sample and record them on the pages that follow. Calculate the average and record this. Record the average result and applicable units. An example has been provided.

Example

Quenchant: Air

Rockwell B Reading 1: 90

Rockwell B Reading 2: 91

Rockwell B Reading 3: 89

Rockwell B Reading Average: 90

Rockwell C Reading 1: N/A

Rockwell C Reading 2: N/A

Rockwell C Reading 3: N/A

Rockwell C Reading Average: N/A

Reported Reading Including Units: 90 RHB

Name _____

Sample 1 Readings

Quenchant: _____

Rockwell B Reading 1: _____

Rockwell B Reading 2: _____

Rockwell B Reading 3: _____

Rockwell B Reading Average: _____

Rockwell C Reading 1: _____

Rockwell C Reading 2: _____

Rockwell C Reading 3: _____

Rockwell C Reading Average: _____

Reported Reading Including Units: _____

Sample 2 Readings

Quenchant: _____

Rockwell B Reading 1: _____

Rockwell B Reading 2: _____

Rockwell B Reading 3: _____

Rockwell B Reading Average: _____

Rockwell C Reading 1: _____

Rockwell C Reading 2: _____

Rockwell C Reading 3: _____

Rockwell C Reading Average: _____

Reported Reading Including Units: _____

Sample 3 Readings

Quenchant: _____

Rockwell B Reading 1: _____

Rockwell B Reading 2: _____

Rockwell B Reading 3: _____

Rockwell B Reading Average: _____

Rockwell C Reading 1: _____

Rockwell C Reading 2: _____

Rockwell C Reading 3: _____

Rockwell C Reading Average: _____

Reported Reading Including Units: _____

Sample 4 Readings

Quenchant: _____

Rockwell B Reading 1: _____

Rockwell B Reading 2: _____

Rockwell B Reading 3: _____

Rockwell B Reading Average: _____

Rockwell C Reading 1: _____

Rockwell C Reading 2: _____

Rockwell C Reading 3: _____

Rockwell C Reading Average: _____

Reported Reading Including Units: _____

6. If any reading on the Rockwell B scale was greater than 100, retest this sample with three readings using the Rockwell C scale.

Review Questions

1. Which sample has the highest average hardness?
   _____
   _____

2. Which sample is the softest?
   _____
   _____

3. Compare the control sample with your treated samples. Is the control hardness close to any of the heat-treated samples?
   _____
   _____

4. If you wanted to make this alloy as strong as possible, what heat treatment would you recommend?
   _____
   _____
   _____

5. If you wanted to make this alloy as soft as possible, so it could be cut more easily, what heat treatment would you recommend?
   _____
   _____
   _____

Name _____ Date _____ Class _____

Determining Hardness and Alloy Type by Spark and File Tests

Introduction
Different steel alloys, particularly those in storage areas, look remarkably similar. While most steels from the mill are marked with alloy numbers, cut sections can lose the marking. Sometimes, a quick check can save a die or saw from being smashed and ruined by an overly hard workpiece. Some forming dies get jammed by metal that is too soft for the design. If a hardness tester is not available, an experienced technician can often tell the approximate hardness using a power grinder or even a hand file. The hardness is a suggestion of the type of steel.

Objective
- Use grinders and files to estimate steel sample hardness and type of steel.

Safety Considerations
- Grinders and hand files produce metal fines you do not want in your eyes. Thus, the need for safety glasses.
- The point where a workpiece meets the grinding wheel is a *pinch point*—where a sleeve or finger could be pulled in and damaged. Keep hands away from the grinder when it is working.

Equipment
- An electric power grinder suitable for softer metals, such as mild steel

> **CAUTION**
> Never use a grinder intended for tool steels to grind mild steels or softer metals. The soft metal gums up the grit surface of the grinder, so it cannot properly sharpen high-strength tool steels. If this mistake is made, the machinist must redress the grinding wheel. Most machine shops have grinders that are dedicated to tool steels. Use those grinders only for tool steels.

- Safety glasses
- At least one pair of leather gloves for holding the samples during grinding
- A vice for holding small samples firmly
- A handheld metal file

Materials
- Samples of high-speed tool steel, mild steel, and cast iron. One of each sample type is marked. Additional samples are probably similar alloys but are not identified. Mark these A, B, C, etc., so later discussions will be clear.

Spark Test Procedure

1. The key to a successful spark test is consistency in holding the specimen relative to the grinding wheel of the power grinder, **Figure 3-1**.

 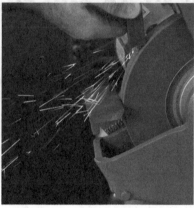 

Goodheart-Willcox Publisher

Figure 3-1. The key to this test is to hold the workpiece against the grinding wheel with the proper angle and pressure. Assuming consistency in grinding technique, the sparks will look different depending on the metal.

There is some skill involved in grinding samples on a grinder. The instructor will illustrate how to safely hold a sample and how to grind it. As the instructor grinds a small portion of each marked sample, stand at a right angle to the stream of sparks so you can see the process better. The sparks are small bits of steel that burn in the air. Sometimes sparks form short paths, sometimes long, straight paths. Sometimes they form short paths that break into multiple paths. The different shape of the spark path is an indication of a different type of steel, and hence the hardness of that steel. Make a note describing the sparks you observe with each of the three known samples.

High-Speed Tool Steel

Spark Observations: _____
_____
_____
_____

Mild Steel

Spark Observations: _____
_____
_____
_____

Cast Iron

Spark Observations: _____
_____
_____
_____

Name _____

2. After observing the sparks created as the instructor grinds a small portion of each of the three known samples, grind a small portion of an identified sample yourself, so you can see the stream of sparks from the grinder's point of view.

3. Take a sample with a letter marking (A, B, C, etc.) and grind a small corner of it. Make a guess of what type of steel it is. Repeat with the other unknown samples, keeping a written note of your identifications. Take note of the shape of the sparks and which of the known samples they most closely match.

Sample A

Spark Observations: _____
_____
_____
_____

Guess of Type of Steel: _____

Sample B

Spark Observations: _____
_____
_____
_____

Guess of Type of Steel: _____

Sample C

Spark Observations: _____
_____
_____
_____

Guess of Type of Steel: _____

Sample D

Spark Observations: _____
_____
_____
_____

Guess of Type of Steel: _____

Review Questions for Spark Tests

1. When other students have tested the same sample of unknown steel, compare notes and see how well you all agree. How do results compare?

   _____
   _____
   _____
   _____
   _____
   _____
   _____

2. Compare your identifications with the instructor's assessments. If you do not agree with the instructor, go back and repeat the spark tests. Look for what the instructor used to make a judgment. How did your results compare?

   _____
   _____
   _____
   _____
   _____
   _____
   _____

File Test Procedure

1. In the hands of an experienced person, the use of a handheld metal file can also suggest different alloys. The instructor will put each of the identified samples in a vice and file off a few fines with the file. Note how the different samples produce different fines.

High-Speed Tool Steel

Description of Fines: _____

_____

Mild Steel

Description of Fines: _____

_____

Cast Iron

Description of Fines: _____

_____

Name _____

2. Put a lettered sample (A, B, C, etc.) in the vice and make your own fines.

Lettered Unknown Sample

Sample Letter: _____

Description of Fines: _____

_____

Guess of Type of Steel: _____

Review Questions for File Test

1. Which known sample is closest to your unknown sample?

   _____
   _____
   _____

2. When other students have tested the same sample of unknown steel, compare notes and see how well you all agree. How did your results compare?

   _____
   _____
   _____
   _____
   _____
   _____
   _____

3. Compare your identifications with the instructor's assessments. How do they compare? If you do not agree with the instructor, discuss what you saw that was different so you can tell the three types of steel apart.

   _____
   _____
   _____
   _____
   _____
   _____
   _____
   _____
   _____

Notes

Name _____ Date _____ Class _____

LAB 4: Hardness Testing of Thin Sheet

Introduction
Testing the hardness of thin sheet can be tricky. It is important to use the correct indenter, force, and scale. If a hardness test sample is too thin, a Rockwell B or C test will report the hardness of the anvil underneath, not the sample. A Rockwell superficial hardness test is used on thin sheet. The hardness test must be chosen to measure the workpiece, not the anvil.

Objective
- Perform Rockwell superficial hardness tests on thin sheets.

Safety Considerations
- Keep hands away from active parts of the hardness testers. The machines move slowly, but with great force.

Equipment
- A Rockwell hardness tester
- A 1/16″ tungsten carbide ball indenter, with weights for 30T and 45T scales for testing thin sheet
- Safety glasses

Materials
- Three samples of cold-rolled brass or annealed steel strip, each about 1″ (2.54 cm) wide by 6″ (15.24 cm) long, and 0.064″ (1.63 mm), 0.032″ (0.81 mm), and 0.020″ (0.51 mm) thick, respectively

Procedure

1. Insert the indenter for thin sheet (1/16″ ball). Use the weight for a 30T scale test. That is, use a test force of 30 kgf.
2. Be sure the test samples are clean and dirt-free. Select a portion of strip that is flat and lies on the anvil without rocking or bowing upward.
3. Make three hardness measurements of each thickness with the previous Rockwell test procedure detailed in Labs 1 and 2: (1) insert sample, (2) raise capstan to indicated preload, (3) release major load, (4) wait, (5) remove major load, (6) read hardness value. Be sure to allow 3/8″ distance between each indent and from any edge. If the sample is too soft, the indenter will report the anvil hardness. Consult the following table to be sure you are using the correct superficial scale and obtaining a hardness that reflects the sample only.

Minimum Thickness Guide for Selection of Scales for 1/16″ (1.59 mm) Ball Indenter*			
Sample Thickness	**15T**	**30T**	**45T**
0.010″ / 0.25 mm	91	…	…
0.012″ / 0.30 mm	86	…	…
0.014″ / 0.36 mm	81	80	…
0.016″ / 0.41 mm	75	72	71
0.018″ / 0.46 mm	68	64	62
0.020″ / 0.51 mm	…	55	53
0.022″ / 0.56 mm	…	45	43
0.024″ / 0.61 mm	…	34	31
0.026″ / 0.66 mm	…	…	18
0.028″ / 0.71 mm	…	…	4
0.030″ / 0.76 mm	…	…	…

*From ASTM E18-22 Standard

Goodheart-Willcox Publisher

Note: Ellipses (…) indicate any hardness is acceptable.

Figure 4-1. Consult this table to be sure you are using the correct superficial scale and obtaining a hardness that reflects only the sample.

> **NOTE**
> Make a superficial hardness reading using the 30T scale and load. If the hardness reading for your thickness of sheet is *less* than the 30T scale allows, use the 15T scale. Following are some examples:

- The 0.018″ sheet reads 63 on the 30T scale. Use the 45T scale.
- The 0.024″ sheet reads 32 on the 30T scale. Use the 15T scale.
- The 0.020″ sheet reads 56 on the 30T scale. OK to report.
- The 0.016″ sheet reads 70 on the 30T scale. Cannot take an accurate Rockwell superficial reading on this sheet with 15T, 30T, or 45T tests.
- The 0.030″ sheet reads 44 on the 30T scale. OK to report.

4. Record your data on the following pages. Convert all superficial hardness values to approximate tensile strength (or another single scale) for comparison.

Name _____

Hardness Test Results

Test 1
Sheet Thickness: 0.064"

Scale Used: _____

Superficial Rockwell Hardness: _____

Estimated Tensile Strength: _____

Test 2
Sheet Thickness: 0.064"

Scale Used: _____

Superficial Rockwell Hardness: _____

Estimated Tensile Strength: _____

Test 3
Sheet Thickness: 0.064"

Scale Used: _____

Superficial Rockwell Hardness: _____

Estimated Tensile Strength: _____

Hardness Average for 0.064" thickness: _____

Test 1
Sheet Thickness: 0.032"

Scale Used: _____

Superficial Rockwell Hardness: _____

Estimated Tensile Strength: _____

Test 2
Sheet Thickness: 0.032"

Scale Used: _____

Superficial Rockwell Hardness: _____

Estimated Tensile Strength: _____

Test 3
Sheet Thickness: 0.032"

Scale Used: _____

Superficial Rockwell Hardness: _____

Estimated Tensile Strength: _____

Hardness Average for 0.032" thickness: _____

Test 1

Sheet Thickness: 0.020"

Scale Used: _____

Superficial Rockwell Hardness: _____

Estimated Tensile Strength: _____

Test 2

Sheet Thickness: 0.020"

Scale Used: _____

Superficial Rockwell Hardness: _____

Estimated Tensile Strength: _____

Test 3

Sheet Thickness: 0.020"

Scale Used: _____

Superficial Rockwell Hardness: _____

Estimated Tensile Strength: _____

Hardness Average for 0.020" thickness: _____

Review Questions

1. Which thickness of sheet has the highest hardness (that is, highest estimated tensile strength)? Which has the lowest?

   _____

   _____

   _____

   _____

   _____

   _____

2. What difference between these pieces of sheet could explain the different strengths?

   _____

   _____

   _____

   _____

   _____

   _____

   _____

Name _____

Discussion Question

1. Keeping in mind that the Rockwell indenter makes an indent in the test sample, explain why the minimum thickness guide gives *minimum hardness* values for each thickness of sheet.

_____
_____
_____
_____
_____
_____
_____
_____

Notes

Name _____ Date _____ Class _____

Hardness Testing of Castings

Introduction
Castings have much coarser microstructures than wrought steel, and some portions of the cast microstructure have much different hardness values than others. Thus, a Brinell indenter, which has a much larger diameter than indenters for Rockwell B and C, will give a more uniform measurement of hardness.

Objectives
- Use a Brinell hardness tester to measure castings.
- Understand why certain hardness test methods yield more accurate results depending on materials being tested.

Safety Considerations
- Keep hands away from the active parts of the hardness testers. The machines move slowly, but with great force.
- Wear safety glasses when tests are being conducted.
- Ensure safe handling of specimens.

Equipment
- A Brinell hardness tester, or a Rockwell hardness tester with a Brinell indenter (10 mm tungsten carbide or hardened steel ball)
- A Rockwell B indenter
- Safety glasses

Materials
- Four or five cast iron samples suitable for hardness testing. Two samples should be gray cast iron from the same source.
- Sandpaper, 300 or finer grit

Procedure

1. Perform three hardness measurements on one gray cast iron sample, record them below, and set this sample aside.

   Reading 1 Brinell Hardness Number (BHN): _____

   Reading 2 Brinell Hardness Number (BHN): _____

   Reading 3 Brinell Hardness Number (BHN): _____

For Testing with a Brinell Hardness Tester:

1. If needed, grind the flat sides of the test samples flat and approximately parallel. Smooth the side to be measured with 300 or finer grit sandpaper.
2. Make hardness measurements with the Brinell hardness tester.
3. Measure the diameter of the indent using the microscope on the tester. The units are millimeters. Calculate the average, and record this.
4. Calculate the Brinell hardness number (BHN) using the average indent diameter and the equation that follows. (You may find it easier to set up this equation using Microsoft Excel on a laptop.)

$$BHN = \frac{P}{\frac{\pi D}{2}(D - \sqrt{D^2 - d^2})}$$

Where:

P = Load in kg (3000 kgf)

D = Diameter of ball (10 mm)

d = Diameter of indentation (in mm)

BHN = Brinell hardness number

For Testing with a Rockwell Hardness Tester:

1. If needed, grind the flat sides of the test samples flat, smooth, and approximately parallel.
2. Use the indenter and settings for Brinell in the Rockwell tester.
3. Perform the hardness test as with other Rockwell tests. Make three measurements, keeping proper spacing, and record them and the average of the three measurements under "Brinell Hardness Measurements."

Brinell Hardness Measurements

Sample 1

Description: _____

Brinell Indent Diameter 1: _____

Brinell Indent Diameter 2: _____

Brinell Indent Diameter 3: _____

Average Indent Diameter: _____

Brinell Hardness Number (use equation): _____

Name _____

Sample 2

Description: _____

Brinell Indent Diameter 1: _____

Brinell Indent Diameter 2: _____

Brinell Indent Diameter 3: _____

Average Indent Diameter: _____

Brinell Hardness Number (use equation): _____

Sample 3

Description: _____

Brinell Indent Diameter 1: _____

Brinell Indent Diameter 2: _____

Brinell Indent Diameter 3: _____

Average Indent Diameter: _____

Brinell Hardness Number (use equation): _____

Sample 4

Description: _____

Brinell Indent Diameter 1: _____

Brinell Indent Diameter 2: _____

Brinell Indent Diameter 3: _____

Average Indent Diameter: _____

Brinell Hardness Number (use equation): _____

Sample 5

Description: _____

Brinell Indent Diameter 1: _____

Brinell Indent Diameter 2: _____

Brinell Indent Diameter 3: _____

Average Indent Diameter: _____

Brinell Hardness Number (use equation): _____

Measurements with Rockwell Hardness Tester

For the gray iron sample that was already tested and set aside at the beginning of the lab:

1. Transfer your initial three hardness readings to the "Repeated Measurements" below. Plan where to put seven more indents. Take seven more Brinell measurements on the gray iron sample you already tested. Interpolate hardness values if the dial falls between two tick marks. Record all 10 measurements under "Repeated Measurements."

For the gray iron sample that has not yet been measured:

1. Plan where to put 10 indents. Reset the Rockwell tester to Rockwell B conditions, and measure the hardness on the B scale 10 times. Measure with as much accuracy as possible. If any reading is over 95, use the Rockwell C scale instead. Record the hardness readings under "Repeated Measurements."

2. Under "Repeated Measurements," convert all BHN and RHB readings to tensile strength (UTS), then record the minimum and maximum strengths under "Summary of Readings." (The UTS number is included in most hardness conversion charts.) Record the *spread*, or difference between the maximum and minimum strengths. Calculate the average. You may also calculate the standard deviation if you like.

Repeated Measurements

Reading 1

Brinell Hardness Number (BHN): _____

BHN Converted to UTS: _____

Rockwell B Hardness (RHB): _____

RHB converted to UTS: _____

Reading 2

Brinell Hardness Number (BHN): _____

BHN Converted to UTS: _____

Rockwell B Hardness (RHB): _____

RHB converted to UTS: _____

Reading 3

Brinell Hardness Number (BHN): _____

BHN Converted to UTS: _____

Rockwell B Hardness (RHB): _____

RHB converted to UTS: _____

Reading 4

Brinell Hardness Number (BHN): _____

BHN Converted to UTS: _____

Rockwell B Hardness (RHB): _____

RHB converted to UTS: _____

Name _____

Reading 5

Brinell Hardness Number (BHN): _____

BHN Converted to UTS: _____

Rockwell B Hardness (RHB): _____

RHB converted to UTS: _____

Reading 6

Brinell Hardness Number (BHN): _____

BHN Converted to UTS: _____

Rockwell B Hardness (RHB): _____

RHB converted to UTS: _____

Reading 7

Brinell Hardness Number (BHN): _____

BHN Converted to UTS: _____

Rockwell B Hardness (RHB): _____

RHB converted to UTS: _____

Reading 8

Brinell Hardness Number (BHN): _____

BHN Converted to UTS: _____

Rockwell B Hardness (RHB): _____

RHB converted to UTS: _____

Reading 9

Brinell Hardness Number (BHN): _____

BHN Converted to UTS: _____

Rockwell B Hardness (RHB): _____

RHB converted to UTS: _____

Reading 10

Brinell Hardness Number (BHN): _____

BHN Converted to UTS: _____

Rockwell B Hardness (RHB): _____

RHB converted to UTS: _____

Summary of Readings

Average BHN: _____

Average UTS (converted from BHN): _____

Average RHB: _____

Average UTS (converted from RHB): _____

Minimum UTS (from BHN): _____

Minimum UTS (from RHB): _____

Maximum UTS (from BHN): _____

Maximum UTS (from RHB): _____

Spread (UTS from BHN): _____

Spread (UTS from RHB): _____

Standard Deviation (UTS from BHN): _____

Standard Deviation (UTS from RHB): _____

Review Questions

1. Consider the procedure and technique you used here and in Lab 1, where Rockwell B and C scales were first used. What differences can you find between techniques?

   _____
   _____
   _____
   _____
   _____
   _____
   _____

2. Review your Repeated Measurements results. We expect that your measurements came out slightly different from one another, even on the same sample. Can you think of possible reasons why this should happen?

   _____
   _____
   _____
   _____
   _____
   _____
   _____

Name _____

3. Which hardness measuring method had a wider spread in results? Can you offer an explanation for this outcome?

   _____
   _____
   _____
   _____
   _____
   _____
   _____
   _____

4. If you were responsible for inspecting and then passing or rejecting gray iron castings from a supplier, which hardness testing method would you prefer, Rockwell B scale or Brinell scale? Why?

   _____
   _____
   _____
   _____
   _____
   _____
   _____
   _____
   _____
   _____

Notes

Name _____ Date _____ Class _____

Tensile Testing of Steel

Introduction
Tensile strength is perhaps the most common measure of steel. It is used in the order and shipping ticket for incoming material to make sure it meets the requirements of the part design, as well as the capabilities of the forming machines.

Objectives
- Determine the tensile strength of three steel samples.
- Understand the material information provided by a tensile test and the purpose of tensile testing.
- Analyze a stress-strain diagram.
- Explain the modulus of elasticity.
- Calculate percent elongation and percent reduction.
- Understand yield and ultimate strength.

Safety Considerations
- A tensile test machine does not move fast, but the forces involved are substantial.
- Wear safety glasses.
- Keep a suitable distance from moving parts, even slowly moving ones.

Equipment
- A tensile test machine
- Extensometer (optional)
- Two small furnaces, capable of 1550°F (843°C) and 600°F (316°C)
- A small steel rule
- A 1" micrometer
- Permanent marker or engraver (optional)

Materials

- Three steel tensile test samples cut from AISI 1045 steel

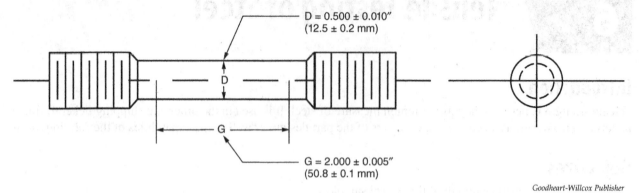

Figure 6-1. Only the D and G dimensions are specified by ASTM. The other dimensions may vary slightly.

Figure 6-2. A typical tensile test sample.

Procedure

1. Place all three tensile samples into a preheated furnace at 1550°F (843°C), spread out so they will heat up evenly and equally. Soak them for 30 minutes.

2. Quench two specimens in water. Place one of these samples into the 600°F (316°C) furnace for one hour. Set the third sample on a bench to cool in air. When cool, clean off the mill scale. Mark the quenched specimen "Q," the tempered specimen "T," and the air-cooled specimen "A."

3. With a permanent marker or an engraver, mark two points 2″ apart on the narrow section of the specimens. If you use an engraver, be sure to mark the sample lightly. (If you have an extensometer, use it instead of these marks.)

Name _____

4. Measure the distance between the marks. The standard length for an extensometer is 1" (2.54 cm). Measure the diameter of each sample near the center of the reduced-diameter length. Record your data.

5. Check the hardness of each sample by selecting an area with a uniform diameter away from the central region of the sample. The large diameter at the end of the sample will work nicely. Use a Rockwell B scale and a cylindrical conversion table. If the reading is above 100, use the C scale. Convert all hardness readings to estimated tensile strength. Record your data.

6. Your instructor will place a sample into the tensile test machine with the extensometer, if available, and "pull" the sample. As the sample stretches, carefully note the stages it goes through—elastic deformation, uniform reduction, local necking, and fracture. (Chapter 4 has more discussion on these stages.)

7. When the sample has broken and the instructor has removed it, examine the break surfaces carefully.

CAUTION
DO NOT touch the broken surfaces together! The damage done this way prevents a failure analyst from getting full information.

Does the sample show signs of a cup-and-cone fracture? Does it resemble a brittle fracture? Record your results.

8. Place the broken pieces almost touching and measure the final distance between the marks. Record your results.

9. Holding the micrometer carefully, measure the diameter of the broken pieces at the failure point. Record your results.

10. Finish filling in the rest of your data and results. Equations for calculations are provided here.

Equations for percent elongation and elastic strain:

$$\text{Strain } e = \frac{\text{Change in length}}{\text{Original length}} \times 100\%$$

$$\text{Elongation} = \frac{\text{Change in length}}{\text{Original length}} \times 100\%$$

$$= \frac{(\text{Final length} - \text{Original length})}{\text{Original length}} \times 100\%$$

Equations for area:

$$\text{Area of test sample cross section} = \pi \times \text{radius}^2 = \pi \times \left(\frac{\text{Diameter}}{2}\right)^2$$

$$\text{Area} = \pi \times \left(\frac{D}{2}\right)^2 = \pi \times \frac{D^2}{4}$$

Equation for percent reduction in area:

$$\text{Percent reduction in area} = \frac{D^2_{\text{initial}} - D^2_{\text{final}}}{D^2_{\text{initial}}} \times 100\%$$

Equation for tensile strength:

$$\text{Stress (psi)} = \frac{\text{Force (lb)}}{\text{Area (in}^2\text{)}}$$

Test Data and Results

As-Quenched Sample

Initial Gauge Length: _____

Final Gauge Length: _____

Percent Elongation: _____

Gauge Length at Yield: _____

Elastic Strain (%): _____

Initial Diameter: _____

Initial Area: _____

Final Diameter: _____

Final Area: _____

Percent Reduction in Area: _____

Type of Failure (Ductile or Brittle): _____

Load at Yield (approx.): _____

Maximum Load: _____

Maximum Tensile Strength: _____

Quenched-and-Tempered Sample

Initial Gauge Length: _____

Final Gauge Length: _____

Percent Elongation: _____

Gauge Length at Yield: _____

Elastic Strain (%): _____

Initial Diameter: _____

Initial Area: _____

Final Diameter: _____

Final Area: _____

Percent Reduction in Area: _____

Type of Failure (Ductile or Brittle): _____

Load at Yield (approx.): _____

Maximum Load: _____

Maximum Tensile Strength: _____

Name _____

Air-Cooled (Normalized) Sample

Initial Gauge Length: _____

Final Gauge Length: _____

Percent Elongation: _____

Gauge Length at Yield: _____

Elastic Strain (%): _____

Initial Diameter: _____

Initial Area: _____

Final Diameter: _____

Final Area: _____

Percent Reduction in Area: _____

Type of Failure (Ductile or Brittle): _____

Load at Yield (approx.): _____

Maximum Load: _____

Maximum Tensile Strength: _____

11. Transfer your tensile strength and elongation data here for a quick comparison of the sample results.

As-Quenched Sample

Percent Elongation: _____

Maximum Tensile Strength: _____

Quenched-and-Tempered Sample

Percent Elongation: _____

Maximum Tensile Strength: _____

Air-Cooled (Normalized) Sample

Percent Elongation: _____

Maximum Tensile Strength: _____

Review Questions

1. The heat treatments given are as-quenched, quenched and tempered, and normalized. Which heat treatment gives the greatest ultimate tensile strength?

   _____

   _____

   _____

2. Which heat treatment gives the greatest elongation?

   _____

   _____

   _____

3. Reduction in area at fracture can be used as an indication of ductility. Which heat treatment is likely to have the greatest ductility?

   _____

   _____

   _____

4. Compare the tensile strength and elongation of samples that received the quenched process to those that received the quenched-and-tempered process. Did you find that tensile strength and elongation rose together, went the opposite directions, or seemingly had no relationship?

   _____

   _____

   _____

   _____

   _____

5. Now, compare the tensile strength and elongation of samples that received the quenched-and-tempered process to those that received the normalized (air-cooled) process. Did you find that tensile strength and elongation rose together, went the opposite directions, or seemingly had no relationship? Are these trends the same as between the quenched and quenched-and-tempered processes compared in question 4?

   _____

   _____

   _____

   _____

   _____

   _____

Name _____ Date _____ Class _____

Metal or Nonmetal?

Introduction
Chapter 5 of the textbook states that if a material is a metal, it has all four properties shared by metals:
- Electrical conductivity
- Thermal conductivity
- Formability, or the ability to deform without cracking
- Reflectivity, or shininess

Some materials have some of these properties, but only metals have all four. Thermal conductivity is closely related to electrical conductivity, so you can skip complex tests for that property in this lab.

Objectives
- Identify metals and nonmetals using the four fundamental physical properties of all metals.
- Understand that some composite and coated items may be tricky to discern.

Safety Considerations
- Wear safety glasses or goggles when smashing things, especially ceramics and rocks, and tap them just hard enough to bend, dent, or break.
- The drinking glass has already been tested for you, so you do not need to break yours. The glass did not bend. You can test your glass for conductivity without breaking it.

Equipment
- Safety glasses or goggles
- A multimeter that can read electrical resistance in ohms
- A hammer and an anvil or hard surface to pound on
- A piece of fine sandpaper (emery board will work)
- Polishing compound and buffing cloth, such as for polishing automobiles (copper or silver polishing compound will also work)

Materials
- Find at least 15 items to test for metallic properties, including the first five from the list of possible test items. Try many different things. Some things are fun to work with because they can behave strangely. The list that follows is incomplete—use your imagination to include other materials.

Possible Test Items:
- A drinking glass made of glass. *Do not break this.* The drinking glass was tested for you, so you do not need to break this one. The glass did not bend. You can test your glass for conductivity without breaking it.
- A ceramic dish (**Note:** Make sure it is okay to break it, first!)
- A wood pencil you can destroy in testing. Break off the eraser end of the pencil and sharpen both ends.
- A piece of copper wire, at least 1′ (30 cm) long
- A flexible refrigerator magnet
- A sheet of paper made of heavy paper stock
- A penny or quarter coin
- A rock 1/2″ to 2″ (1 to 5 cm) in diameter. A piece chipped off a boulder is great. A small pebble will also work.
- A celery stalk, beet leaf stem, or other produce
- A piece of plywood or processed wood flooring at least 6″ (15 cm) long
- A cast iron fry pan, griddle, or skillet
- A plastic cup or empty plastic soda bottle
- A TV screen wall mount
- A garden rake, hoe, or shovel

> **NOTE**
> Make sure you are testing what an object is made of, not the coating on it. Clean the surface of each item to reach and test the material underneath the surface. Dry organic material may give a different answer than wet organic material. Anything with a wet surface may conduct electricity on the wet surface. Firmly press the point of the current probe into the surface to test conductivity.

Procedure

1. Gather 15 or more items from what you find or from the above list. List the items you plan to test under "Materials Testing Results."
2. From your experience, you already have a good idea of what is made of metal. In a laboratory, this is called a hypothesis, or an educated guess. For each item you've chosen, next to "Hypothesis," mark "Y" if you guess the item is made of metal and "N" if you think it is a nonmetal.
3. Perform the ductility test (Test A), the polish test (Test B), and the conductivity test (Test C) on each item.

Test A: Deforms
The hypothesis for this test is "something that dents or bends might be a metal." If it shatters, it is not a metal.

1. Bend or dent each item until it clearly bends, dents, or fractures.
2. For each item, next to "Deforms," enter a "Y" for anything that dented or bent and an "N" for anything that shattered.

Test B: Polishes
The hypothesis is "something that can be polished might be a metal."

> **NOTE**
> Sanding the surface might damage the item. Be sure this damage is OK before proceeding.

1. With the fine sandpaper or emery board, rub one of the surfaces you can easily get to. After the sandpaper has removed any surface paint or coating, use the polishing compound and buffing cloth to polish the surface a little more.
2. For each item, next to "Polishes," enter a "Y" for good polish and an "N" for not polishable. Be careful not to confuse a special finish, such as epoxy or glaze, with the actual material of the item.

Name _____

Test C: Conducts

The hypothesis is "something that conducts electricity might be a metal."

1. Put the two leads against each test piece, about 1″–2″ (2–5 cm) apart. Be sure they make good contact with the test piece. If the resistance drops below 10 ohms, the material is conductive. If the ohmmeter reads one megohm or greater, it is definitely not conductive. When using batteries and an LED indicator, if the light turns on, the object is definitely conductive.

2. For each item, next to "Conducts," enter a "Y" for things that conduct electricity and an "N" for any nonconductors.

Materials Testing Results

Item 1: drinking glass

Hypothesis (metal, Y/N): _____

Test A, Deforms (Y/N): _____

Test B, Polishes (Y/N): _____

Test C, Conducts (Y/N): _____

Metal (Y/N): _____

Hypothesis Confirmed (Y/N): _____

Item 2: ceramic dish

Hypothesis (metal, Y/N): _____

Test A, Deforms (Y/N): _____

Test B, Polishes (Y/N): _____

Test C, Conducts (Y/N): _____

Metal (Y/N): _____

Hypothesis Confirmed (Y/N): _____

Item 3: pencil

Hypothesis (metal, Y/N): _____

Test A, Deforms (Y/N): _____

Test B, Polishes (Y/N): _____

Test C, Conducts (Y/N): _____

Metal (Y/N): _____

Hypothesis Confirmed (Y/N): _____

Item 4: copper wire

Hypothesis (metal, Y/N): _____

Test A, Deforms (Y/N): _____

Test B, Polishes (Y/N): _____

Test C, Conducts (Y/N): _____

Metal (Y/N): _____

Hypothesis Confirmed (Y/N): _____

Item 5: flexible magnet

Hypothesis (metal, Y/N): _____

Test A, Deforms (Y/N): _____

Test B, Polishes (Y/N): _____

Test C, Conducts (Y/N): _____

Metal (Y/N): _____

Hypothesis Confirmed (Y/N): _____

Item 6:

_____

Hypothesis (metal, Y/N): _____

Test A, Deforms (Y/N): _____

Test B, Polishes (Y/N): _____

Test C, Conducts (Y/N): _____

Metal (Y/N): _____

Hypothesis Confirmed (Y/N): _____

Item 7:

_____

Hypothesis (metal, Y/N): _____

Test A, Deforms (Y/N): _____

Test B, Polishes (Y/N): _____

Test C, Conducts (Y/N): _____

Metal (Y/N): _____

Hypothesis Confirmed (Y/N): _____

Name _____

Item 8:

_____

Hypothesis (metal, Y/N): _____

Test A, Deforms (Y/N): _____

Test B, Polishes (Y/N): _____

Test C, Conducts (Y/N): _____

Metal (Y/N): _____

Hypothesis Confirmed (Y/N): _____

Item 9:

_____

Hypothesis (metal, Y/N): _____

Test A, Deforms (Y/N): _____

Test B, Polishes (Y/N): _____

Test C, Conducts (Y/N): _____

Metal (Y/N): _____

Hypothesis Confirmed (Y/N): _____

Item 10:

_____

Hypothesis (metal, Y/N): _____

Test A, Deforms (Y/N): _____

Test B, Polishes (Y/N): _____

Test C, Conducts (Y/N): _____

Metal (Y/N): _____

Hypothesis Confirmed (Y/N): _____

Item 11:

_____

Hypothesis (metal, Y/N): _____

Test A, Deforms (Y/N): _____

Test B, Polishes (Y/N): _____

Test C, Conducts (Y/N): _____

Metal (Y/N): _____

Hypothesis Confirmed (Y/N): _____

Item 12:

_____

Hypothesis (metal, Y/N): _____

Test A, Deforms (Y/N): _____

Test B, Polishes (Y/N): _____

Test C, Conducts (Y/N): _____

Metal (Y/N): _____

Hypothesis Confirmed (Y/N): _____

Item 13:

_____

Hypothesis (metal, Y/N): _____

Test A, Deforms (Y/N): _____

Test B, Polishes (Y/N): _____

Test C, Conducts (Y/N): _____

Metal (Y/N): _____

Hypothesis Confirmed (Y/N): _____

Item 14:

_____

Hypothesis (metal, Y/N): _____

Test A, Deforms (Y/N): _____

Test B, Polishes (Y/N): _____

Test C, Conducts (Y/N): _____

Metal (Y/N): _____

Hypothesis Confirmed (Y/N): _____

Item 15:

_____

Hypothesis (metal, Y/N): _____

Test A, Deforms (Y/N): _____

Test B, Polishes (Y/N): _____

Test C, Conducts (Y/N): _____

Metal (Y/N): _____

Hypothesis Confirmed (Y/N): _____

Name _____

Item 16:

_____

Hypothesis (metal, Y/N): _____

Test A, Deforms (Y/N): _____

Test B, Polishes (Y/N): _____

Test C, Conducts (Y/N): _____

Metal (Y/N): _____

Hypothesis Confirmed (Y/N): _____

Item 17:

_____

Hypothesis (metal, Y/N): _____

Test A, Deforms (Y/N): _____

Test B, Polishes (Y/N): _____

Test C, Conducts (Y/N): _____

Metal (Y/N): _____

Hypothesis Confirmed (Y/N): _____

Item 18:

_____

Hypothesis (metal, Y/N): _____

Test A, Deforms (Y/N): _____

Test B, Polishes (Y/N): _____

Test C, Conducts (Y/N): _____

Metal (Y/N): _____

Hypothesis Confirmed (Y/N): _____

Item 19:

_____

Hypothesis (metal, Y/N): _____

Test A, Deforms (Y/N): _____

Test B, Polishes (Y/N): _____

Test C, Conducts (Y/N): _____

Metal (Y/N): _____

Hypothesis Confirmed (Y/N): _____

Item 20:

_____

Hypothesis (metal, Y/N): _____

Test A, Deforms (Y/N): _____

Test B, Polishes (Y/N): _____

Test C, Conducts (Y/N): _____

Metal (Y/N): _____

Hypothesis Confirmed (Y/N): _____

Results

1. Look at the results from your tested items.
2. For each item, if any test (Deforms, Polishes, Conducts) is an "N" result, write an "N" next to "Metal." The item is *not* metal.
3. If *all* tests (Deforms, Polishes, Conducts) for an item have a "Y" result, write a "Y" next to "Metal." The item is metal.
4. Write a "Y" or "N" next to "Hypothesis" for each item to indicate either yes, your hypothesis is confirmed, or no, your hypothesis was incorrect.

Review Questions

1. Is there anything in your list of confirmed metals you did not expect to be metal? List any items.

   _____
   _____
   _____
   _____

2. Is there anything among your tested items you expected to be made of metal but is not? List any items.

   _____
   _____
   _____
   _____

3. How do you explain the flexible refrigerator magnet, which is magnetic yet not conductive? Is it made of metal?

   _____
   _____
   _____
   _____
   _____

Name _____

4. Did you find a material whose properties you cannot easily explain? (Some high-strength plastics can fool people.) List any items.

_____
_____
_____
_____
_____

5. Things made of two or more materials, such as tennis rackets reinforced with graphite fiber, may indicate electrical conductivity even though only one of the components is conductive. Did you find any composites that yielded strange results? Can you find a reasonable explanation for the behavior of these items?

_____
_____
_____
_____
_____
_____
_____
_____
_____

Notes

Name _____ Date _____ Class _____

LAB 8
Metallurgical Specimen Preparation

Introduction
With preparation, we can see many metallurgical structures under an optical microscope between 50X and 1000X magnification.

Objective
- Prepare specimens for microscope viewing.

Safety Considerations
- Safety glasses shall be worn by everyone in the metallurgical preparation area at all times. This includes even visitors who are simply "passing through," as well as other visitors. You are empowered to remind any neglectful classmate or other person of this rule, should you see anyone without safety glasses.
- During preparation, samples can fly out of control—keep an eye out whenever someone is polishing a sample.
- Etchants can damage skin and eyes. If you get any etchant on your skin, flush it off with water and alert your instructor.

Equipment
- Safety glasses
- Tongs
- Microscope suitable for 50X to 500X magnification
- A metallurgical lab, such as in **Figure 8-1**, includes equipment for mounting a sample, grinding and polishing, and etching the cross section. Once these steps are accomplished, the metallurgical structure of the cross section can be viewed under a microscope.

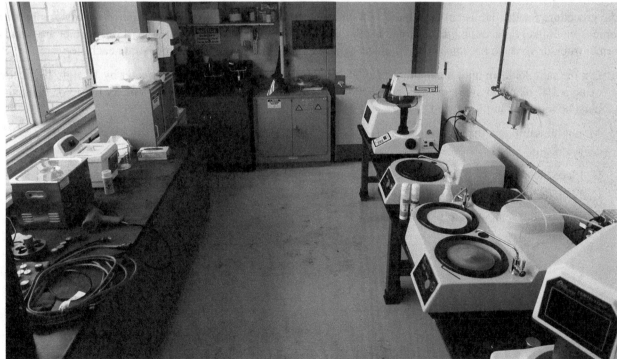

Goodheart-Willcox Publisher

Figure 8-1. A typical metallurgical lab, with equipment for mounting a sample for microscopic viewing on the left and polishing wheels on the right.

Sample Mounting Equipment

- Thermosetting resin is heated under pressure to make an easily handled, uniform shape holding the sample. The resin can be a phenolic or epoxy.

Grinding and Polishing Equipment

- Belt sander to grind off the outer portion of the sample
- Wet sander and progressively smaller sandpaper: 240 (sand until flat), 320, 400, 600, 800, and 1200 grit
- Circular polishing wheel pad with 1 µm diamond slurry
- Another circular polishing pad with 0.05 µm alumina slurry

NOTE
These two circular polishing pads must use only the assigned polishing slurry.

Etching Equipment

- Fume hood
- Etchants
- Water rinse

Materials

Consumable materials for the laboratory are supplied.

- Polishing pads, for example: 1 micron diamond or alumina, 1 micron or smaller final polishing alumina diamond, colloidal silica/diamond intermediate
- Slurries, for example: 1 µm low-viscosity polycrystalline diamond suspension, 0.05 µm alpha alumina slurry, non-crystallizing colloidal silica
- 2% nital etchant
- Samples from previous labs, such as Lab 1, supplied by your instructor

Procedure

The procedure for this lab is "in the hands"—in other words, you must do it to learn it. Good metallographers often come from unlikely places and professional backgrounds. (The authors are aware of at least one former artist and one former truck driver who became excellent metallographers!) The instructor will demonstrate each step.

1. Cut the sample down until it can fit into the hot mount press, approximately 1″ diameter by 1″ high. If you are using a 3/4″ diameter slug, you may polish it directly. If it is an odd shape or much smaller than 3/4″ diameter, it must be mounted in plastic first.

2. Mount the sample in thermosetting resin:

    A. Preheat the press.

    B. Place the sample face down on the piston and surround it with resin powder.

    C. Lock the chamber and apply pressure for the time specified.

    D. Release the sample (carefully—it is hot!).

Name _____

3. Grind the exposed side of the sample flat with a belt sander. Use the standard method for holding the sample, as demonstrated by your instructor:

   A. Use the thumb, forefinger, and second finger together to make a sound three-point contact with the sample.

   B. Stand so that the grinding surface is moving directly away from you. Do not stand to one side of the grinder because that makes it harder to hold the sample firmly.

   C. Hold the sample so the sander cuts in only one direction. Grind off 1/16″ of material, or until all signs of machining marks are gone. The sander will leave parallel lines of tiny ridges on the surface.

The grinding and polishing in steps 4–7 should be for one minute at 200 rpm, with water for coolant and about 10 pounds of force applied.

4. Grind off the tiny ridges using the wet sander and 240 grit sandpaper. Grind/polish the sample 90° from the previous cut, so totally new and smaller ridges are developed.

5. Turn the sample 90° and polish it using the 320 grit sandpaper.

6. Repeat step 5 using 400 grit sandpaper, then repeat using 600 grit, then 800 grit, and finally 1200 grit sandpaper.

7. Polish the surface using the circular wheel with the 1 μm diamond paste. Again, turn the sample 90° from the previous polish. Push the sample slowly against the direction of the wheel, keeping it wet with the slurry or water. It will take about 2 minutes at 200 rpm.

8. Rinse all slurry off the sample and your fingers thoroughly, then turn the sample 90° again and polish on the 0.05 μm alumina wheel at 100 rpm for 30 seconds to reach the final polish.

9. Rinse the sample and your fingers to remove any remaining slurry.

10. Etch the sample:

    A. Hold the sample in a pair of tongs by the round sides.

    B. Dip the sample face into the nital etchant, putting it in one side first so no air bells get trapped under it. Hold it there for a few seconds—between 5 and 30 seconds, depending on how deep you want to etch. Move the sample around gently as you hold it in the solution.

    C. Remove the sample and rinse it off completely under the gently running water. A steel sample will look slightly brownish. Dry the sample by using a short rinse in alcohol and then waving it in the air until dry. Wiping with a cotton swab is OK.

> **CAUTION**
>
> **Etching**
> All etching is to be done under the hood, and the hood fan must be on while metallography is performed. Nital is a solution of nitric acid in propanol. Always keep the petri dish for the etching solutions well under the hood. Nitric acid can make a nasty burn on organic material, such as your hand. Keep water running gently in the sink the entire time etchants are out and available. Avoid splashing rinse water. Rinse samples thoroughly before taking them into the microscope room. Microscopes do not like acids any more than your skin does.
>
> If you do accidentally touch an etching solution, simply dip the contacted area under the running water. Nitric acid in reasonable concentrations will leave only a small yellow area. With the working solutions used in this lab, you may not notice anything. However, be aware that some metallurgical laboratory solutions can cause considerable short- and long-term damage. Some are explosive under specific conditions, and some etch organic materials until they neutralize on bone.

11. Take the sample to the microscope room, and examine it under the microscope.

Review Questions

1. Why are specimens mounted?

2. Why do the specimen surfaces need to be flat for microscope viewing?

3. What is the purpose of etching?

4. If your sample is mill run or normalized AISI 1020 to 1045, you will see islands of pearlite in a clear field of ferrite. AISI 1080 will be completely pearlite. If you can compare different steel grades, can you see differences in the amount of pearlite? Describe your observations.

Name _____ Date _____ Class _____

Ductile-to-Brittle Transition Temperature of Steel

Introduction

Above room temperature, plain carbon steels fail in a ductile manner. Deformation occurs around the failure site, and the fracture surface has a characteristic dull gray appearance. Significant energy is required to cause failure.

At temperatures below 32°F (0°C), the plain carbon steel fails in a brittle manner. Little or no deformation occurs near the fracture, and the fracture surface has a characteristic sparkle. The energy required for fracture is low; that is, the steel has low impact toughness. Brittle failure occurs at higher temperatures for higher-carbon steel. That is, the temperature of the transition in fracture mode is higher for higher-carbon steel.

Objective

- Determine the brittle-to-ductile transition temperature of steel, and compare the behavior of plain carbon and stainless steel.

Safety Considerations

Charpy impact tester: The Charpy impact tester is probably the most dangerous piece of equipment in the laboratory. The impact tester head swings down with a lot of energy. Stay out of the line of fire of the impact tester and take the following precautions:

- Do not get hands or other body parts in the way of the hammer.
- Stay out of the enclosed area with the hammer lifted or in the ready position. Only one person should be at the control, and everyone else should stay well outside the test area.
- Whenever the weight is in the upper position, the person who lifted it must remain with one shoulder resting under it so that if the latch should release, the person will catch it. Do not leave this position until allowed by the person placing the sample.
- The person who places the sample in the fixture is the only person who may trip the release.
- Everyone should stay away from the area the broken samples go toward unless retrieving them.
- Do not try to catch the hammer. Use the brake to slow it, or wait for the swinging to stop.

Liquid nitrogen: Liquid nitrogen is very cold and can cause frostbite and burns on contact. Even the vapor near liquid nitrogen is cold and can cause skin damage. A room-temperature sample causes vigorous boiling when first placed into liquid nitrogen, and the liquid can spatter. Take the following precautions concerning liquid nitrogen:

- Use tongs to place samples into the cold liquid.
- Use safety glasses, a face shield, and gloves when handling the liquid nitrogen.
- Be careful with anything that comes into contact with the liquid nitrogen; the container, tongs, and specimens will cause damage to skin.

Boiling water:

- Wear safety glasses and gloves when handling the boiling water.

Equipment

- Universal impact tester, **Figure 9-1**

A *Goodheart-Willcox Publisher* **B** *Goodheart-Willcox Publisher*
Figure 9-1. A—On a universal impact tester, the weight at the end of the arm swings down to strike a sample at the bottom. B—The scale on the machine reports the energy absorbed by the sample in foot-pounds or joules.

- A hot plate and a glass beaker for heating water
- Beakers for holding ice water and liquid nitrogen
- Low-power stereo microscope (10–20X magnification)
- Tongs
- Safety glasses (your own pair)
- Face shield
- Gloves

Materials

- Four standard Charpy V-notch specimens—10 mm x 10 mm x 40 mm—of each alloy, made of:
  AISI 1018 or A36 steel
  AISI 1045 steel (A higher carbon content may be substituted.)
  AA 6061 aluminum
  AISI 304 stainless steel
- Liquid nitrogen
- Water and ice

Procedure

1. Mark the Charpy samples with a marker to indicate the alloy for each.

Name _____

Notes on Use of Impact Tester

1. Ensure the needle of the slide bar ft-lb scale is all the way back to 0 ft-lb. When the swing arm is lifted into the primed position, you can hear the latch mechanism click. If you now let it back down, the latch will hold it. However, keep one shoulder under it to ensure you are not surprised.

2. Use tongs to pick up each sample and place it in the fixture with the notch facing away from the impact point, **Figure 9-2**. Center the sample against the stop blocks on the side away from the swing arm head.

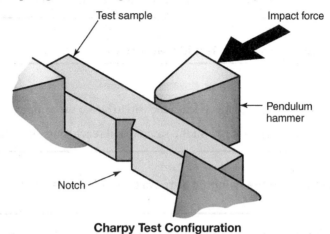

Figure 9-2. The standard Charpy test configuration is illustrated.

3. Step back after placing the sample. When everyone is clear of the work area, reach out and press the trip button on the machine. Everyone should be in front, on the dial side of the impact testing machinet, **Figure 9-3**. Do not stand behind the machine or in the path of the hammer at any distance.

Figure 9-3. A—The hammer is swinging down toward the sample. B—The hammer has swung down through the sample.

4. After the anvil has reached its full height on the other side of the sample and starts to swing back, then—and only then—press the brake button. This activates a magnetic clutch to stop the swing arm motion.

Room Temperature Impact Test

1. Select one sample of each alloy, and place these samples in a small group separate from the rest of the samples.
2. Using the tongs, place one sample in the Charpy impact tester on the test shelf with the notch facing away from the hammer.
3. Clear away from the device and move the hammer into its top position. Make sure the hammer is latched in position before letting go of it. Always have one person hold the hammer in addition to using the latch.

> **SAFETY**
> Make sure no one is in range of the hammer. **This is a safety issue!**

4. Place the dial at 120.
5. The person who placed the sample should now release the trip to strike the sample.
6. Apply the brake to stop the hammer after it has hit the sample and reached its full height on the other side.

> **SAFETY**
> **Do not attempt to manually stop the swinging arm.** Wait for the hammer to stop before retrieving the broken samples.

7. Check the slide gauge on the impact tester and record the absorbed impact energy under "Impact Test Data."
8. Lay the broken pieces out on a large bench for examination, label them with the test temperature, and record the temperatures under "Impact Test Data."
9. Using the Fracture Appearance chart as a guide, **Figure 9-4**, record the approximate percentage of brittle and ductile fracture failure under "Impact Test Data."
10. Repeat the impact test with each of the three remaining room temperature samples.

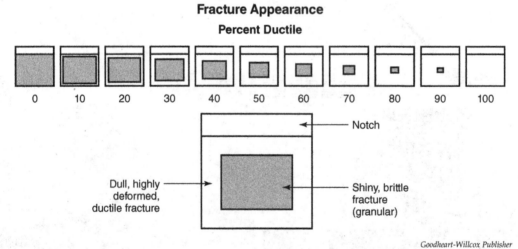

Figure 9-4. Use this chart to determine the absorbed impact energy for each test.

Ice Water Impact Test

1. Fill a pan with ice and put one Charpy sample of each alloy into it. Allow 9 minutes for the samples to cool to ice water temperature.
2. Remove one sample with tongs and place it in the Charpy impact tester on the test shelf with the notch facing away from the hammer. Move quickly but deliberately to keep the temperature true.

Name _____

3. Clear away from the device and move the hammer into its top position. Make sure the hammer is latched in position before letting go of it. Always have one person hold the hammer in addition to using the latch.

> **SAFETY**
> Make sure no one is in range of the hammer. **This is a safety issue!**

4. Place the dial at 120.
5. The person who placed the sample should now release the trip to strike the sample.
6. Apply the brake to stop the hammer after it has hit the sample.

> **SAFETY**
> **Do not attempt to manually stop the swinging arm.** Wait for the hammer to stop before retrieving the broken samples.

7. Check the slide gauge on the impact tester and record the absorbed impact energy with the rest of your impact test data.
8. Lay the broken pieces out on a large bench for examination, label them with the test temperature, and record the temperatures with the rest of your data.
9. Using the Fracture Appearance chart as a guide, **Figure 9-4**, record the approximate percentage of brittle and ductile fracture failure with the rest of your impact test data.
10. Repeat the impact test with each of the three remaining cold samples.

Boiling Water Impact Test

1. Boil water in a saucepan or beaker. Put one Charpy sample for each alloy into the boiling water for 9 minutes.
2. Remove one sample with tongs and place it in the impact tester on the test shelf with the notch facing away from the hammer. Move quickly but deliberately to keep the temperature true.
3. Clear away from the device and move the hammer into the top position. Make sure the hammer is latched in position before letting go of it. Always have one person hold the hammer in addition to using the latch.

> **SAFETY**
> Make sure no one is in range of the hammer. **This is a safety issue!**

4. Place the dial at 120.
5. The person who placed the sample should now release the trip to strike the sample.
6. Apply the brake to stop the hammer after it has hit the sample.

> **SAFETY**
> **Do not attempt to manually stop the swinging arm.** Wait for the hammer to stop before retrieving the broken samples.

7. Record the absorbed impact energy from the impact tester slide gauge with the rest of your impact test data.
8. Lay the broken pieces out on a large bench for examination, label them with the test temperature, and record the temperatures.
9. Using the Fracture Appearance chart as a guide, record the approximate percentage of brittle and ductile fracture failure for the sample.
10. Repeat the impact test with each of the remaining boiling water samples and record your results.

Liquid Nitrogen Impact Test

> **CAUTION**
> The temperature of the liquid nitrogen is –320°F (–195.8°C).

1. Pour some liquid nitrogen into a beaker. Put one Charpy sample for each alloy into the liquid nitrogen. Wait until each sample becomes quiet in the liquid before removing it.

2. Remove one sample with tongs and place it in the impact tester on the test shelf with the notch facing away from the hammer. Move quickly but deliberately to keep the temperature true.

3. Clear away from the device and move the hammer into the top position. Make sure the hammer is latched in position before letting go of it. Always have one person hold the hammer in addition to using the latch.

> **SAFETY**
> Make sure no one is in range of the hammer. **This is a safety issue!**

4. Place the dial at 120.

5. The person who placed the sample should now release the trip to strike the sample.

6. Apply the brake to stop the hammer after it has hit the sample.

> **SAFETY**
> **Do not attempt to manually stop the swinging arm.** Wait for the hammer to stop before retrieving the broken samples.

7. Record the absorbed impact energy from the impact tester slide gauge under "Impact Test Data."

8. Lay the broken pieces out on a large bench for examination, label them with the test temperature, and record the temperatures with your other data.

9. Using the Fracture Appearance chart as a guide, record the approximate percentage of brittle and ductile fracture failure for the sample.

10. Repeat the impact test with each of the remaining liquid nitrogen samples and record your results.

Name _____

Impact Test Data

Test 1

Material: _____

Temperature: _____

Energy Absorbed (ft-lb): _____

Fracture Appearance (% ductile): _____

Notes/Remarks: _____

Test 2

Material: _____

Temperature: _____

Energy Absorbed (ft-lb): _____

Fracture Appearance (% ductile): _____

Notes/Remarks: _____

Test 3

Material: _____

Temperature: _____

Energy Absorbed (ft-lb): _____

Fracture Appearance (% ductile): _____

Notes/Remarks: _____

Test 4

Material: _____

Temperature: _____

Energy Absorbed (ft-lb): _____

Fracture Appearance (% ductile): _____

Notes/Remarks: _____

Test 5

Material: _____

Temperature: _____

Energy Absorbed (ft-lb): _____

Fracture Appearance (% ductile): _____

Notes/Remarks: _____

Test 6

Material: _____

Temperature: _____

Energy Absorbed (ft-lb): _____

Fracture Appearance (% ductile): _____

Notes/Remarks: _____

Test 7

Material: _____

Temperature: _____

Energy Absorbed (ft-lb): _____

Fracture Appearance (% ductile): _____

Notes/Remarks: _____

Test 8

Material: _____

Temperature: _____

Energy Absorbed (ft-lb): _____

Fracture Appearance (% ductile): _____

Notes/Remarks: _____

Test 9

Material: _____

Temperature: _____

Energy Absorbed (ft-lb): _____

Fracture Appearance (% ductile): _____

Notes/Remarks: _____

Test 10

Material: _____

Temperature: _____

Energy Absorbed (ft-lb): _____

Fracture Appearance (% ductile): _____

Notes/Remarks: _____

Name _____

Test 11

Material: _____

Temperature: _____

Energy Absorbed (ft-lb): _____

Fracture Appearance (% ductile): _____

Notes/Remarks: _____

Test 12

Material: _____

Temperature: _____

Energy Absorbed (ft-lb): _____

Fracture Appearance (% ductile): _____

Notes/Remarks: _____

Test 13

Material: _____

Temperature: _____

Energy Absorbed (ft-lb): _____

Fracture Appearance (% ductile): _____

Notes/Remarks: _____

Test 14

Material: _____

Temperature: _____

Energy Absorbed (ft-lb): _____

Fracture Appearance (% ductile): _____

Notes/Remarks: _____

Test 15

Material: _____

Temperature: _____

Energy Absorbed (ft-lb): _____

Fracture Appearance (% ductile): _____

Notes/Remarks: _____

Test 16

Material: _____

Temperature: _____

Energy Absorbed (ft-lb): _____

Fracture Appearance (% ductile): _____

Notes/Remarks: _____

Analysis

- Examine selected fracture surfaces directly (without magnification) and under a low-power stereo microscope at 10–20X magnification. Use the Fracture Appearance chart, **Figure 9-4**, to estimate the percent ductile fraction of each fracture.
- Select one alloy you tested and compare the energy absorbed (ft-lb) at each temperature for that metal.

Review Questions

1. Consider the alloy you selected for an analysis of energy absorbed at each temperature. Does temperature make any noticeable difference for this alloy? Explain.

   _____
   _____
   _____
   _____
   _____
   _____
   _____

2. Consider the rest of your data from the impact tests. Based on your data, how does alloy content alter a metal's impact toughness?

   _____
   _____
   _____
   _____
   _____
   _____
   _____

Name _____ Date _____ Class _____

LAB 10 Hardenability of Steel

Introduction
Chapter 12 discusses heat-treating heavy sections of steel. It is simply impossible to quench heavy sections fast enough to obtain maximum hardness and strength through the entire part. Certain alloy additions modify the response to quenching so that the hardest material extends deeper into the heavy section. In this lab, you will quench and measure some samples to see these alloy effects for yourself. An ASTM procedure called a Jominy test or end-quench test defines the significant details of the test.

Objective
- Learn how to measure hardenability and explore which alloys harden deeper into a large block of steel.

Safety Considerations
- You will be working with a hot furnace at 1650°F (900°C), so the usual gloves and tongs are needed.
- Be aware that quenching hot metal into water may cause steam and flying hot water.
- Be sure to maintain an exit path from the furnace room.
- If you should drop a sample while it is hot, do not try to pick it up immediately. Be sure you are out of harm's way first, then safely return it to the furnace to reheat.

Equipment
- A furnace set at 1650°F (900°C). It must be large enough to hold three Jominy test samples vertically at one time.
- A Jominy test stand, **Figure 10-1**. It has a small water tank and a pump to force water through a nozzle to make a vertical jet of water. Immediately above the nozzle is a support to hold the test samples.

A   *Goodheart-Willcox Publisher*   B   *Goodheart-Willcox Publisher*

Figure 10-1. A—A typical Jominy test stand. B—Top of the Jominy test stand, with a hole in the center of a support bar where the sample is placed for the quench test.

- Rockwell hardness tester. A special fixture on the Rockwell hardness tester will hold the cylindrical Jominy samples.
- Safety goggles or glasses
- Gloves and tongs for handling hot samples

Materials

- Three end-quench samples: one each of AISI 1040, 8620, and 4340 or 4140. These are 1″ in diameter by approximately 4″ long, with a flange at one end, as shown in **Figure 10-2**. There are three stainless steel cups that each hold a specimen. (Yes, they are stainless steel, even though they are heavily corroded.)

Goodheart-Willcox Publisher
Figure 10-2. A standard Jominy end-quench test sample.

Procedure

1. Preheat the furnace to 1650°F (900°C).
2. Put each Jominy sample into the stainless steel cup, then place each sample into the furnace upright.
3. Soak for 45 minutes.
4. While the samples are soaking, set up the end-quench apparatus. Check that the water level is high enough and refill it if necessary. Set the water flow valve so it produces a jet of water 2.5″ above the end of the nozzle.
5. Practice a few "dry runs." Place an old sample into the apparatus and turn on the water. Remember, you have only a few seconds to get the sample out of the furnace, put it in the apparatus, and start the water. In the past, students have found it best to set the flow valve while the pump is running, then turn the water on and off with the pump switch. Do *not* start the water until the samples are fully in place.
6. When the samples have soaked enough, remove one from the furnace and set the cup and sample upright on the bench. Lift the sample out by the flange and place it into the Jominy quench machine. Continue to quench it in the apparatus until the end away from the water has cooled to about 500°F (260°C). This will take about 10 minutes.
7. If the sample has not yet been stamped with the alloy designation, mark that on it, as well your class and the time.
8. Repeat the Jominy end-quench test (steps 6 and 7) for the next two samples.
9. Turn the samples over to your instructor, who will arrange to have two parallel flats ground along the length of the samples to a depth of 0.015″ (0.38 mm). The instructor or tech will take care to avoid heating the samples during grinding, since this could soften the harder portions.

Name _____

10. When the samples are returned to you, take hardness readings. These are done using the special support fixture with the Rockwell hardness tester, **Figure 10-3**. One turn of the lead screw on this fixture is 1/16″. Measure the hardness starting from the quenched end, getting no closer than 1/16″ from the actual end. Measure the first inch in 1/16″ increments, and measure the second inch in 1/8″ increments.

Goodheart-Willcox Publisher

Figure 10-3. Jominy end-quench test sample in the support fixture ready to be hardness tested on the Rockwell tester.

Jominy Hardness Results

Alloy 1 Hardness (HRC) Measurements

0″ from End: _____

1/4″ from End: _____

1/2″ from End: _____

3/4″ from End: _____

1″ from End: _____

1 1/4″ from End: _____

1 1/2″ from End: _____

1 3/4″ from End: _____

2″ from End: _____

Alloy 2 Hardness (HRC) Measurements

0″ from End: _____

1/4″ from End: _____

1/2″ from End: _____

3/4″ from End: _____

1″ from End: _____

1 1/4″ from End: _____

1 1/2″ from End: _____

1 3/4″ from End: _____

2″ from End: _____

Alloy 3 Hardness (HRC) Measurements

0" from End: _____

1/4" from End: _____

1/2" from End: _____

3/4" from End: _____

1" from End: _____

1 1/4" from End: _____

1 1/2" from End: _____

1 3/4" from End: _____

2" from End: _____

Review Question

1. Examine your Jominy hardness results. Describe any noteworthy observations here.

_____

_____

_____

_____

_____

_____

_____

Name _____ Date _____ Class _____

Tempering Martensite

Introduction
As you certainly know by the time you complete Chapter 10 of the textbook, martensite is a very hard, brittle phase in steel that is created when the metal is heated above 1341°F (727°C) and then quenched rapidly into water. It is also *metastable*, meaning it can change into softer, more ductile microstructures when aged at a moderate temperature.

Objective
- Learn how martensite hardness changes when it is aged at different temperatures.

Safety Considerations
- You are processing samples in hot furnaces, so you will use gloves and tongs to handle the samples going into or out of the furnaces. Also, before entering a confined space with furnaces, always check how you would leave the area quickly should that be necessary. Quenching hot samples into water may cause hot water splashes, so be prepared for this and use a face shield.

Equipment
- One or multiple furnaces
- Sandpaper and holder, 300–600 grit (600 grit preferred)
- Rockwell hardness tester with RHB and RHC indenters
- Gloves
- Tongs
- Safety glasses and face shield

Materials
- Five samples of plain carbon steel, each 3/4″ diameter and 3/4″ long. The steel may be AISI 1045 or AISI 1095.
- Water suitable for quenching samples

Procedure
1. Measure the hardness of all samples using the Rockwell B scale and record your results under "Rockwell B Hardness at Room Temperature."
2. Soak all specimens in a 1650°F (900°C) furnace for 30 minutes, spacing them out for uniform heating.
3. Quench all specimens in water, one at a time.
4. Mark specimens by number on a round side using a stamp or engraver.
5. Clean and measure the hardness of quenched specimens using the Rockwell C scale, and record your results under "Rockwell C Hardness of Quenched Samples."

Copyright Goodheart-Willcox Co., Inc.
May not be reproduced or posted to a publicly accessible website.

Rockwell B Hardness at Room Temperature

Sample 1 Hardness, HRB: _____

Sample 2 Hardness, HRB: _____

Sample 3 Hardness, HRB: _____

Sample 4 Hardness, HRB: _____

Sample 5 Hardness, HRB: _____

Rockwell C Hardness of Quenched Samples

Sample 1 Hardness, HRC: _____

Sample 2 Hardness, HRC: _____

Sample 3 Hardness, HRC: _____

Sample 4 Hardness, HRC: _____

Sample 5 Hardness, HRC: _____

Procedure with One Furnace

If only one furnace is available, use the following steps:

1. Turn the furnace completely off and leave the furnace door open to cool it faster. When the furnace is under 200°F, set it to 200°F.

2. When the furnace has stabilized to 200°F (i.e., within 4°F of set point), place all the specimens in the furnace.

3. Soak the specimens for 10 minutes.

4. Using the tongs, quench one sample in water.

5. Clean the flat surfaces and measure the sample's hardness.

6. Record the result under "Hardness and Soak Temperature Results."

7. Raise the furnace temperature to 400°F, wait until the furnace is within 5°F of that set temperature, then soak the samples for 10 more minutes.

8. Quench one sample, clean it, and measure and record its hardness.

9. Raise the furnace temperature to 600°F, wait until the furnace is within 10° of that set temperature, then soak the remaining samples for 10 more minutes.

10. Quench one sample, clean it, and measure and record its hardness.

11. Raise the furnace temperature to 800°F, wait until the furnace is within 10° of that set temperature, then soak the remaining samples for 10 more minutes.

12. Quench one sample, clean it, and measure and record its hardness.

13. Raise the furnace temperature to 1000°F, wait until the furnace is within 10° of that set temperature, then soak the remaining sample for 10 more minutes.

14. Quench the last sample, clean it, and measure and record its hardness.

Name _____

Procedure with Multiple Furnaces

If multiple furnaces are available, preset them to the temperatures being tested to make more efficient use of lab time, and use the following steps:

1. Preset one furnace to 200°F. When the temperature is stable (i.e., within 4°F of set point), place all the specimens in the furnace. Keep the samples spaced apart so they will heat more uniformly.

2. Soak the samples for 10 minutes.

3. Using the tongs, quench one sample in water.

4. Clean the flat surfaces of the sample, and measure its hardness.

5. Record the result under "Hardness and Soak Temperature Results."

6. Preset a furnace to 400°F. When the temperature is stable, move all remaining samples into this furnace, and soak them for 10 minutes. Keep the samples spaced apart.

7. Using the tongs, quench one sample, clean it, and measure and record its hardness.

8. Preset a furnace to 600°F. When the temperature is stable, move all remaining samples into this furnace, and soak them for 10 minutes. Keep the samples spaced apart.

9. Using the tongs, quench one sample, clean it, and measure and record its hardness.

10. Preset a furnace to 800°F. When the temperature is stable, move all remaining samples into this furnace, and soak them for 10 minutes. Keep the samples spaced apart.

11. Using the tongs, quench one sample, clean it, and measure and record its hardness.

12. Preset a furnace to 1000°F. When the temperature is stable, move the remaining sample into this furnace, and soak it for 10 minutes.

13. Using the tongs, quench the last sample, clean it, and measure and record its hardness.

Hardness and Soak Temperature Results

Maximum Soak Temperature: 200°F

Sample 1 Hardness, HRC: _____

Maximum Soak Temperature: 400°F

Sample 2 Hardness, HRC: _____

Maximum Soak Temperature: 600°F

Sample 3 Hardness, HRC: _____

Maximum Soak Temperature: 800°F

Sample 4 Hardness, HRC: _____

Maximum Soak Temperature: 1000°F

Sample 5 Hardness, HRC: _____

Review Questions

1. What happened to the hardness of samples as the maximum soak temperature increased?

   _____
   _____
   _____
   _____
   _____

2. What do you expect would happen to the hardness if you soaked a fully martensitic sample at 1000°F for 1 hour?

   _____
   _____
   _____
   _____
   _____

3. What do you expect would happen to the hardness if you soaked a fully martensitic sample at 1200°F for 10 minutes?

   _____
   _____
   _____
   _____
   _____

Name _____ Date _____ Class _____

Effects of Annealing on Cold-Worked Brass

Introduction
When deformation such as cold-rolling or forging of a workpiece occurs, we know that dislocation tangles are created and dislocation motion is more difficult. We see this as a higher-strength, or harder, metal. We also know that if the workpiece is annealed, the dislocations can be removed and the metal hardness, or strength, drops.

Objective
- Explore the effects of annealing on cold-worked brass.

Safety Considerations
- Even at "only" 1100°F (593°C), samples coming out of a furnace are still too hot to touch. Use tongs, gloves, and safety glasses. Avoid splashes of hot water.

Equipment
- One furnace set to 1100°F (593°C)
- Rockwell hardness tester set to the F scale
- Timers (cell phones work well)
- Safety glasses
- Gloves
- Tongs

Materials
- Eleven small pieces of cold-rolled brass, each about 1″ square and 1/4″ thick
- Water to quench the samples, located very close to the furnace

Procedure

1. Place five of the brass samples in the furnace, toward the back, spread around the bottom and not on top of one another. Start a timer as soon as you have them in place. The first sample in this group will come out at 21 minutes. Someone in the class should keep track of this time.

2. Close the furnace for 1 minute. Then place six samples spread around the front of the furnace, and close the door quickly. Watch the time carefully because some of these samples will come out very soon. For this group, aim for the desired time, within 10 seconds.

3. At 1 minute: Remove one sample from this group at the front of the furnace. Keep the furnace closed as much as possible to maintain the temperature. Quench the sample in water, mark it for time, and make three hardness readings. Report the readings under "Sample Hardness Records." Report the soak time to the nearest second using decimal fractions of a minute, from the moment it went into the furnace until it hit the water.

4. At 2 minutes: Remove a second sample from this group. Quench it, mark it, and make three hardness readings. Report the readings in the table. Report time to the nearest second using decimal fractions of a minute.

5. At 3 minutes: Remove another sample from this group. Quench it, mark it, and make three hardness readings. Record the readings and the time to the nearest second using decimal fractions of a minute.

6. At 5 minutes: Remove another sample, quench it, mark it, and make three hardness readings. Record the readings and the time to the nearest second using decimal fractions of a minute.

7. At 8 minutes: With another sample, repeat the process from step 6.

8. At 13 minutes: Remove the last sample of the group at the front of the furnace. Repeat the process from step 6.

9. Now for the samples at the back of the furnace. For this group, aim for the desired time, within 1 minute.

10. At 21 minutes: Remove a sample from the furnace. Process it like the previous samples, and record your results. Consider your results so far, and answer Review Question 2 at the end of this lab.

11. At 35 minutes: Remove, process, and record results for another sample.

12. At 60 minutes (1 hour): Remove, process, and record results for another sample.

13. At 100 minutes (1 hour and 40 minutes): Remove, process, and record results for another sample.

14. At 160 minutes (2 hours and 40 minutes): Remove, process, and record results for the last sample. (Yes, this is a long wait. Perhaps one person can stay after class to record results for this last sample.)

Name _____

Sample Hardness Records

Sample 1 Desired Soak Time: 1.0 minute

Actual Soak Time: _____

Soak Time in Log_{10} Minutes: _____

Hardness Reading (HRF) 1: _____

Hardness Reading (HRF) 2: _____

Hardness Reading (HRF) 3: _____

HRF Hardness Average: _____

Sample 2 Desired Soak Time: 2.0 minutes

Actual Soak Time: _____

Soak Time in Log_{10} Minutes: _____

Hardness Reading (HRF) 1: _____

Hardness Reading (HRF) 2: _____

Hardness Reading (HRF) 3: _____

HRF Hardness Average: _____

Sample 3 Desired Soak Time: 3.0 minutes

Actual Soak Time: _____

Soak Time in Log_{10} Minutes: _____

Hardness Reading (HRF) 1: _____

Hardness Reading (HRF) 2: _____

Hardness Reading (HRF) 3: _____

HRF Hardness Average: _____

Sample 4 Desired Soak Time: 5.0 minutes

Actual Soak Time: _____

Soak Time in Log_{10} Minutes: _____

Hardness Reading (HRF) 1: _____

Hardness Reading (HRF) 2: _____

Hardness Reading (HRF) 3: _____

HRF Hardness Average: _____

Sample 5 Desired Soak Time: 8.0 minutes

Actual Soak Time: _____

Soak Time in Log_{10} Minutes: _____

Hardness Reading (HRF) 1: _____

Hardness Reading (HRF) 2: _____

Hardness Reading (HRF) 3: _____

HRF Hardness Average: _____

Sample 6 Desired Soak Time: 13.0 minutes

Actual Soak Time: _____

Soak Time in Log_{10} Minutes: _____

Hardness Reading (HRF) 1: _____

Hardness Reading (HRF) 2: _____

Hardness Reading (HRF) 3: _____

HRF Hardness Average: _____

Sample 7 Desired Soak Time: 21.0 minutes

Actual Soak Time: _____

Soak Time in Log_{10} Minutes: _____

Hardness Reading (HRF) 1: _____

Hardness Reading (HRF) 2: _____

Hardness Reading (HRF) 3: _____

HRF Hardness Average: _____

Sample 8 Desired Soak Time: 35.0 minutes

Actual Soak Time: _____

Soak Time in Log_{10} Minutes: _____

Hardness Reading (HRF) 1: _____

Hardness Reading (HRF) 2: _____

Hardness Reading (HRF) 3: _____

HRF Hardness Average: _____

Name _____

Sample 9 Desired Soak Time: 60.0 minutes (1 hour)

Actual Soak Time: _____

Soak Time in $\log_{10}$ Minutes: _____

Hardness Reading (HRF) 1: _____

Hardness Reading (HRF) 2: _____

Hardness Reading (HRF) 3: _____

HRF Hardness Average: _____

Sample 10 Desired Soak Time: 100.0 minutes (1 hour and 40 minutes)

Actual Soak Time: _____

Soak Time in $\log_{10}$ Minutes: _____

Hardness Reading (HRF) 1: _____

Hardness Reading (HRF) 2: _____

Hardness Reading (HRF) 3: _____

HRF Hardness Average: _____

Sample 11 Desired Soak Time: 160.0 minutes (2 hours and 40 minutes)

Actual Soak Time: _____

Soak Time in $\log_{10}$ Minutes: _____

Hardness Reading (HRF) 1: _____

Hardness Reading (HRF) 2: _____

Hardness Reading (HRF) 3: _____

HRF Hardness Average: _____

Analysis

1. Start the analysis while you are waiting for the samples that take more time.
2. Calculate the soak times in $\log_{10}$ minutes and add the results to your records.

Review Questions

1. You were requested to place the brass samples spread around the bottom of the furnace and not on top of one another. Why is this significant?

   _____
   _____
   _____
   _____
   _____

2. After you record results for the samples from the first 30 minutes of the lab, discuss what hardness values you expect to see in the remaining samples.

   _____
   _____
   _____
   _____
   _____

3. Collectively, if you were to construct a chart from your data, consider why a plot using $\log_{10}$ would be more informative than a simple time series chart.

   _____
   _____
   _____
   _____
   _____

Name _____ Date _____ Class _____

Determining Strength of a Weld

Introduction
This lab will demonstrate the effect of welding on tensile strength. Depending on students' skill with welding and sample preparation, results could vary greatly.

Objective
- Determine the strength of a weld you made yourself.

Safety Considerations
- Follow standard procedures for welding and use all standard PPE. Conduct welding in an approved hot-work area.

Equipment
- Tensile test machine
- Welding equipment and PPE
- Metal power grinder or milling machine
- Metal file
- 1″ micrometer or calipers
- Machinist's 6″ rule
- Safety glasses

Materials
- Strips of low-carbon steel, 1″ wide by 4″ long by 1/4″ or 3/8″ thick. At least two strips for each student are required.

Procedure

1. Using whichever welding process your instructor deems convenient, weld two strips of steel together (refer to **Figure 13-1**). It is best practice to choose a filler metal or electrode with a higher tensile strength than the base material for a sound weld.

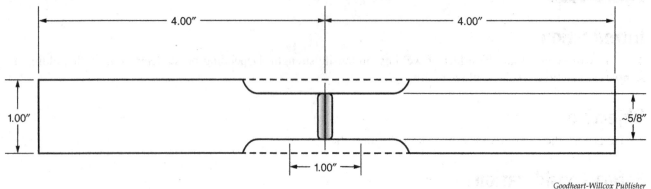

Figure 13-1. Use this figure as a reference for completing your welds.

2. Grind across the weld region to make a narrow "waist," per **Figure 13-1**. Keep the waist width as uniform as possible. A milling machine will do the work best, if you have access to one. Remove any burrs on the edges.
3. Measure the width across the weld and the thickness of the sheet. Record your results under "Weld Records."
4. If the grinding is uneven, attempt to measure at the narrowest portion. If there are uneven areas, smooth them out with a file or sandpaper rather than leave sharp indents.
5. Measure and record the width and thickness of one 4" piece of steel. Put this piece in the tensile test machine and determine the ultimate tensile strength. You will need to calculate the UTS using the equations that follow. This is the strength of your original strip.

For area (A):

$$\text{Cross section area} = \text{Thickness} \times \text{Width}$$

$$A_{Section} = T \times W$$

For tensile strength (UTS):

$$\text{Tensile strength} = \frac{\text{Max load}}{\text{Area}}$$

$$UTS = \frac{\text{Load}}{A_{Section}}$$

6. Put the welded sample into the tensile test machine, and pull it to find the ultimate tensile strength. Watch carefully (with safety glasses on!) to see where the initial yielding occurs. It could be in the weld, in a portion of the heat-affected zone, or in the original metal.

Name _____

Weld Records

Parent Metal: _____

Welder: _____

Thickness: _____

Width: _____

Cross Section: _____

Max Load: _____

UTS: _____

Initial Yield Pont: _____

Failure Location: _____

Review Questions

1. Where was the point of initial yielding?

   _____
   _____
   _____

2. Where was the location of the failure?

   _____
   _____
   _____

3. Compare the UTS of your weld with the strength of the unwelded strip. Which sample has higher strength?

   _____
   _____
   _____
   _____
   _____
   _____

Discussion Question

1. Suppose this weld were in a go-cart and, in a test, it failed at a lower strength than the parent metal. How would you feel about being the driver in the next race? Explain.

_____

_____

_____

_____

_____

_____

Name _____ Date _____ Class _____

Age Hardening of Aluminum

Introduction
Precipitation hardening is a major strengthening mechanism in metals. It is used in HSLA (high-strength low-alloy) steels, certain copper alloys, some aluminum alloys, and especially in aerospace applications. It is even used in lead battery alloys. But age hardening can go too far—a situation called overaging.

The 6061 aluminum alloy in this lab is used in fixtures, window frames, and the like. This is the kind of aluminum that will be available at a hardware store and is common in many terrestrial applications.

Objective
- Learn how heat-treatable aluminum hardens by natural aging (at room temperature) and by artificial aging (time at temperature).

Safety Considerations
- Even furnaces for aging aluminum can burn skin. This temperature is more than double the temperature of an oven at home. The furnaces must be respected, and use of proper PPE is required.
- Hot aluminum looks like cool aluminum, as opposed to steel, which glows in various colors at elevated temperatures. Therefore, before you pick up a sample, pass your hand a few inches over the top of it if you have any doubts as to its temperature. Use gloves and tongs while handling hot samples.
- When working in tight corners, such as in many furnace rooms, always check first to make sure you have a clear exit path should you need one. There is no reason to anticipate that you will need to clear out in a hurry, but you should have plenty of working space so that if a sample does jump toward you, you will have somewhere to go.

Equipment
- Two or more furnaces: one furnace set to 1050°F (566°C) and the second set to 450°F (232°C)
- Rockwell hardness tester, set to the F scale
- Refrigerator with freezer section
- Storage location for this class
- Proper PPE (gloves, tongs, safety glasses)

Materials
- Sixteen aluminum 6061 slugs, 1/2" diameter by 1/2" length
- Water for quenching samples

Procedure
1. Soak (solution heat-treat) all 16 samples at 1050°F (566°C) for 20 minutes at temperature. At the end of the solution heat-treat time, quench all the samples in water, one at a time.

2. When you quench a sample, take it out of the furnace with a pair of tongs, thrust it into the water rapidly, and swirl it around to make sure the water makes good contact with the entire sample. This is important with this alloy.

3. Immediately after quenching, take two Rockwell hardness readings on each of two of the samples using the F scale. Record your results under "Sample Aging and Hardness Records." Set aside four samples that will be used in the "Natural Aging" part of the procedure.

Artificial Aging Tests

1. Put 10 samples into the 450°F (232°C) furnace, spaced out so they heat evenly.
2. At 2 minutes: Remove one sample, quench it, and measure the hardness. Take two readings on the sample, and record the aging time at 450°F (232°C) under "Artificial Aging" in the "Sample Aging and Hardness Records," along with the hardness results.
3. At 3 minutes: Remove, quench, measure, and record the results for another sample.
4. Process another sample in the same manner at each of the following times:
   A. At 8 minutes
   B. At 13 minutes
   C. At 21 minutes
   D. At 35 minutes
   E. At 60 minutes (1 hour)
   F. At 100 minutes (1 hour and 40 minutes)
   G. At 160 minutes (2 hours and 40 minutes), if possible
   H. At over 8 hours (overnight), if possible

Natural Aging

1. Use two as-quenched samples set aside after the first step of the lab procedure. Measure these once per day for one week. (Yes, you may skip Saturday and Sunday if no one can get into the lab.) Record the results under "Natural Aging at Room Temperature" in the "Sample Aging and Hardness Records."
2. Place two samples in the freezer section of a refrigerator, and measure these again one week later, during your next lab. Record the results under "Natural Aging at Freezing Temperature" in the "Sample Aging and Hardness Records."

Sample Aging and Hardness Records

Sample 1:

Aging Temperature: Room Temperature (_____)

Time at Temperature (minutes): _____

Log_{10} time: _____

Hardness Reading (HRF) 1: _____

Hardness Reading (HRF) 2: _____

Average Hardness Reading (HRF): _____

Sample 2:

Aging Temperature: Room Temperature (_____)

Time at Temperature (minutes): _____

Log_{10} time: _____

Hardness Reading (HRF) 1: _____

Hardness Reading (HRF) 2: _____

Average Hardness Reading (HRF): _____

Name _____

Artificial Aging

Sample 1:

Time at Temperature: 2 minutes

Aging Temperature: 450°F

Log_{10} time: _____

Hardness Reading (HRF) 1: _____

Hardness Reading (HRF) 2: _____

Average Hardness Reading (HRF): _____

Sample 2:

Time at Temperature: 3 minutes

Aging Temperature: 450°F

Log_{10} time: _____

Hardness Reading (HRF) 1: _____

Hardness Reading (HRF) 2: _____

Average Hardness Reading (HRF): _____

Sample 3:

Time at Temperature: 8 minutes

Aging Temperature: 450°F

Log_{10} time: _____

Hardness Reading (HRF) 1: _____

Hardness Reading (HRF) 2: _____

Average Hardness Reading (HRF): _____

Sample 4:

Time at Temperature: 13 minutes

Aging Temperature: 450°F

Log_{10} time: _____

Hardness Reading (HRF) 1: _____

Hardness Reading (HRF) 2: _____

Average Hardness Reading (HRF): _____

Sample 5:

Time at Temperature: 21 minutes

Aging Temperature: 450°F

Log_{10} time: _____

Hardness Reading (HRF) 1: _____

Hardness Reading (HRF) 2: _____

Average Hardness Reading (HRF): _____

Sample 6:

Time at Temperature: 35 minutes

Aging Temperature: 450°F

Log_{10} time: _____

Hardness Reading (HRF) 1: _____

Hardness Reading (HRF) 2: _____

Average Hardness Reading (HRF): _____

Sample 7:

Time at Temperature: 60 minutes (1 hour)

Aging Temperature: 450°F

Log_{10} time: _____

Hardness Reading (HRF) 1: _____

Hardness Reading (HRF) 2: _____

Average Hardness Reading (HRF): _____

Sample 8:

Time at Temperature: 100 minutes (1 hour, 40 minutes)

Aging Temperature: 450°F

Log_{10} time: _____

Hardness Reading (HRF) 1: _____

Hardness Reading (HRF) 2: _____

Average Hardness Reading (HRF): _____

Sample 9:

Time at Temperature: 160 minutes (2 hours, 40 minutes)

Aging Temperature: 450°F

Log_{10} time: _____

Hardness Reading (HRF) 1: _____

Hardness Reading (HRF) 2: _____

Average Hardness Reading (HRF): _____

Sample 10:

Time at Temperature: 8+ hours (overnight)

Aging Temperature: 450°F

Log_{10} time: _____

Hardness Reading (HRF) 1: _____

Hardness Reading (HRF) 2: _____

Average Hardness Reading (HRF): _____

Name _____

Natural Aging at Room Temperature

Room Temperature: _____

Time at Temperature: 24 hours (1 day)

Log_{10} time: _____

Sample 1:

Hardness Reading (HRF) 1: _____

Hardness Reading (HRF) 2: _____

Average Hardness Reading (HRF): _____

Sample 2:

Hardness Reading (HRF) 1: _____

Hardness Reading (HRF) 2: _____

Average Hardness Reading (HRF): _____

Time at Temperature: 48 hours (2 days)

Log_{10} time: _____

Sample 1:

Hardness Reading (HRF) 1: _____

Hardness Reading (HRF) 2: _____

Average Hardness Reading (HRF): _____

Sample 2:

Hardness Reading (HRF) 1: _____

Hardness Reading (HRF) 2: _____

Average Hardness Reading (HRF): _____

Time at Temperature: 72 hours (3 days)

Log_{10} time: _____

Sample 1:

Hardness Reading (HRF) 1: _____

Hardness Reading (HRF) 2: _____

Average Hardness Reading (HRF): _____

Sample 2:

Hardness Reading (HRF) 1: _____

Hardness Reading (HRF) 2: _____

Average Hardness Reading (HRF): _____

Copyright Goodheart-Willcox Co., Inc.
May not be reproduced or posted to a publicly accessible website.

Time at Temperature: 96 hours (4 days)

Log_{10} time: _____

Sample 1:

Hardness Reading (HRF) 1: _____

Hardness Reading (HRF) 2: _____

Average Hardness Reading (HRF): _____

Sample 2:

Hardness Reading (HRF) 1: _____

Hardness Reading (HRF) 2: _____

Average Hardness Reading (HRF): _____

Time at Temperature: 120 hours (5 days)

Log_{10} time: _____

Sample 1:

Hardness Reading (HRF) 1: _____

Hardness Reading (HRF) 2: _____

Average Hardness Reading (HRF): _____

Sample 2:

Hardness Reading (HRF) 1: _____

Hardness Reading (HRF) 2: _____

Average Hardness Reading (HRF): _____

Time at Temperature: 144 hours (6 days)

Log_{10} time: _____

Log_{10} time: _____

Sample 1:

Hardness Reading (HRF) 1: _____

Hardness Reading (HRF) 2: _____

Average Hardness Reading (HRF): _____

Sample 2:

Hardness Reading (HRF) 1: _____

Hardness Reading (HRF) 2: _____

Average Hardness Reading (HRF): _____

Name _____

Time at Temperature: 168 hours (7 days)

Log₁₀ time: _____

Sample 1:

Hardness Reading (HRF) 1: _____

Hardness Reading (HRF) 2: _____

Average Hardness Reading (HRF): _____

Sample 2:

Hardness Reading (HRF) 1: _____

Hardness Reading (HRF) 2: _____

Average Hardness Reading (HRF): _____

Natural Aging at Freezing Temperature

Aging Temperature: 0°F

Time at Temperature: 7 days (1 week)

Log₁₀ time: _____

Sample 1:

Hardness Reading (HRF) 1: _____

Hardness Reading (HRF) 2: _____

Average Hardness Reading (HRF): _____

Sample 2:

Hardness Reading (HRF) 1: _____

Hardness Reading (HRF) 2: _____

Average Hardness Reading (HRF): _____

Analysis

1. Take a close look at your records and analyze your hardness results for any trends and takeaways.

2. You may want to highlight your average hardness results to make it easier to analyze and compare them for broad trends.

Review Questions

1. Is there a trend in the freezer sample measurements? Explain.

   _____

   _____

   _____

   _____

   _____

2. Is there a trend in the room-temperature sample measurements? Explain.

   _____
   _____
   _____
   _____
   _____

3. Is there a trend in the 450°F sample measurements? Is there a trend in the room-temperature measurements in the first 20 minutes? Explain.

   _____
   _____
   _____
   _____
   _____
   _____
   _____
   _____

4. At what temperature do the samples change hardness the fastest? The slowest?

   _____
   _____
   _____
   _____

5. Aluminum rivets for airplanes are stored in freezers at −40°F until they are installed. Why might that be a regular procedure?

   _____
   _____
   _____
   _____
   _____

6. Normally, aluminum is artificially aged to peak strength, then returned to room temperature, where it remains at the peak strength. What are parts that have been aged three times longer than the time for peak strength called?

   _____
   _____
   _____
   _____

Name _____ Date _____ Class _____

Pack Carburizing of Steel

Introduction
Packing steel in carbon material and soaking it for a time will increase the carbon concentration near the surface. The steel can then be heated again and quenched to increase the surface strength to above the level of the original metal composition.

Objectives
- Surface harden steel by diffusing carbon into it.
- Recognize the effect of carbon content on steel hardness.

Safety Considerations
- Be sure to use proper PPE, especially safety glasses, because crushing charcoal briquettes makes bits of flying debris.
- Wear a mask over the nose and mouth when crushing briquettes. Crushing creates charcoal dust.

Equipment
- Single oven
- Metal tray with lid (electrical conduit box works)
- Optional: indent marking tools to number samples
- Wire brush
- Mallet for crushing
- Safety glasses, mask, and other appropriate PPE

Materials
- Six pieces of 1018 steel, about 1" square by 1/2" thick
- Half bag of charcoal briquets
- Sandpaper and sanding block, 300 to 600 grit (600 preferred)

Procedure

1. Set the furnace to 1700°F (927°C).
2. Smash charcoal into 1/4″ or smaller pieces. Wear safety glasses and a mask over your nose and mouth, and do this step outside or with good hood ventilation.
3. Place charcoal about 1/2″ deep in the tray.
4. If you have a steel indenting and stamping kit, number the samples. Your marks must withstand soaking in the furnace.
5. Place four steel pieces in a tray, spaced at least 1/2″ apart, 1″ from all sides, and 1″ below the top of the tray.
6. On a sheet of paper, sketch the layout of the samples in the tray. Later, you will want to recover some samples with minimum disruption to the others.
7. Cover the pieces in charcoal on all sides.
8. Tamp the charcoal down gently but firmly onto the buried samples.
9. Put the cover on the tray, and place the tray in the oven.
10. Put two steel specimens in the oven, beside the tray.
11. Mark the oven as "Test in progress—do not disturb," and keep the sample layout sheet nearby.

If the furnace is dedicated to this project for 24 hours, use the following steps:

> **NOTE**
> If you do not have access to the lab and furnace at these exact hours, adjust them as needed with the help of your instructor.

   A. At 2 hours: Open the tray and carefully remove one sample. Keep the other samples covered in charcoal. Allow the sample to air-cool on a workbench and mark it to ID the soak time. Be sure your marks can withstand another heating cycle. Record the specimen number and furnace soak time under "Pack Carburization Results."
   B. At 10 hours: Remove another sample. Air-cool and ID it.
   C. At 18 hours: Remove another sample. Air-cool and ID it.
   D. At 24 hours: Remove the last tray sample. Air-cool and ID it.
   E. Remove the two samples from outside the tray and air-cool them.

If the furnace is available for only 6 hours, use the following steps:

   A. At 1 hour: Open the tray and carefully remove one sample. Keep the other samples covered in charcoal. Allow the sample to air-cool on a workbench and mark it to ID the soak time. Be sure your marks can withstand another heating cycle. Record the specimen number and furnace soak time under "Pack Carburization Results."
   B. At 2.5 hours: Remove another sample. Air-cool and ID it.
   C. At 4 hours: Remove another sample. Air-cool and ID it.
   D. At 6 hours: Remove the last sample. Air-cool and ID it.
   E. Remove the two samples from outside the tray and air-cool them.

12. Clean all samples with a wire brush or equivalent. Measure the hardness of each sample using a Rockwell superficial 45T scale. (The Vickers scale will also work if you clean the surfaces well.) Record your results under "Pack Carburization Results."
13. With the oven still at 1700°F, place all samples in the oven for 15 minutes, well spaced out.
14. After 15 minutes, quench each sample in water, one at a time. Minimize the time between oven and water.

Name _____

Pack Carburization Results

1018 Steel Sample 1

Specimen Number: _____

Time in Charcoal at 1700°F: _____

Hardness before Treatment: _____

Hardness after Treatment: _____

1018 Steel Sample 2

Specimen Number: _____

Time in Charcoal at 1700°F: _____

Hardness before Treatment: _____

Hardness after Treatment: _____

1018 Steel Sample 3

Specimen Number: _____

Time in Charcoal at 1700°F: _____

Hardness before Treatment: _____

Hardness after Treatment: _____

1018 Steel Sample 4

Specimen Number: _____

Time in Charcoal at 1700°F: _____

Hardness before Treatment: _____

Hardness after Treatment: _____

1018 Steel Sample 5

Specimen Number: _____

Time in Charcoal at 1700°F: _____

Hardness before Treatment: _____

Hardness after Treatment: _____

1018 Steel Sample 6

Specimen Number: _____

Time in Charcoal at 1700°F: _____

Hardness before Treatment: _____

Hardness after Treatment: _____

Review Questions

1. Compare the as-heat-treated samples that were soaked outside the pack with the samples that received the longest charcoal exposure time. Which have a higher hardness?

   _____
   _____
   _____
   _____

2. Did the superficial hardness of the pack-carburized samples increase with exposure time?

   _____
   _____
   _____
   _____

3. If the superficial hardness did increase with exposure time, why do you suppose that happened?

   _____
   _____
   _____
   _____

4. Why do you think a Rockwell superficial hardness test is called for in this lab?

   _____
   _____
   _____
   _____

5. What are some examples of products that are case-hardened? Why is this superior to thru-hardening in some applications?

   _____
   _____
   _____
   _____
   _____
   _____

Name _____ Date _____ Class _____

LAB 16: Strain-Hardening Copper Wire

Introduction
When wire is drawn through a die, the wire with the new, smaller diameter must be strong enough to pull the larger wire through. Fortunately, the drawing process work-hardens the metal, so the wire coming out is stronger than the wire going in. This allows the metal to be drawn through a small reduction.

Objective
- Draw copper wire and understand the impact of strain-hardening and annealing on the wire-drawing process.

Safety Considerations
- Wear your safety glasses for this lab.
- When wire under tension breaks or exits a drawing die, it can whip around uncontrollably. Stay clear of possible wire breaks and watch from a cautious distance while it is being drawn.

Equipment
- Tensile test machine with sample grips for 0.040″–0.025″ wire
- Drawing dies for reducing wire diameter, with diameters of 0.040″, 0.035″, 0.032″, 0.028″, and 0.025″.

> **NOTE**
> Metric dies may have slightly different dimensions than the values converted from inches. Use the dimensions given by the dies for calculations and reporting.

- Wire cutters
- Vice grips or pliers
- Safety glasses
- Annealing furnace

Materials
- At least 14′ of 0.040″ diameter copper wire
- Masking tape
- Permanent marker
- Lubricating oil or grease

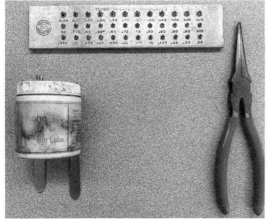
A   *Goodheart-Willcox Publisher*

B   *Goodheart-Willcox Publisher*

Figure 16-1. A—Top: Die for reducing wire diameter; Left: Paste lubricant; Right: Pliers. B—Annealed copper wire on a metal spool.

Procedure

> **NOTE**
> Throughout this work, be careful to not put a crease or kink in the copper wire. A kink in the wire will ruin any measurement.

1. Cut 25″ of 0.040″ diameter copper wire. Carefully put a piece of masking tape on it, and mark the tape for diameter and label it as annealed or cold-drawn. This piece should be marked 0.040″, as received. ("As received" means you have not yet done any annealing or drawing.)

2. "Point" one end of the remaining wire, lubricate the wire with the lubricant, and draw it through the 0.035″ die. Apply lubrication by pulling the wire gently between lubricant-coated fingers. Grip the tip of the wire with the pliers, being careful to hold the wire straight on with the pliers. If you bend the pliers sideways, you could well make a notch in the copper wire, causing it to break off before you pull it.

3. Cut two 25″ lengths of wire, mark them both for diameter, and mark one as cold-drawn. The second piece will be annealed before testing.

4. Repeat steps 2 and 3 using the 0.032″ die.

5. Repeat steps 2 and 3, drawing the wire with a smaller die each time, until the smallest die has been used.

6. Remove the masking tape, taking careful note of the wires' locations, and put the samples to be annealed in a preheated furnace at 1100°F (593°C) for 15 minutes. (The masking tape will burn off and create unnecessary fumes if placed in the furnace.)

7. Perform tensile tests on all the wire samples. Be careful when putting on the clamps—if a clamp makes a notch in the wire and it fails near a grip, the measured load is suspect. It will be less than the true strength of the wire.

8. As you pull samples, record the measured values and your calculations under "Wire Sample Data." Necessary equations follow:

Area:

$$A = \frac{\pi D^2}{4}$$

Percent cold-worked:

$$\% \, CW = \frac{(A_{initial} - A_{final})}{A_{initial}} \times 100$$

Ultimate tensile strength:

$$UTS = \frac{Max \, load}{A}$$

Percent elongation:

$$\% \, elongation = \frac{(Length_{final} - Length_{initial})}{Length_{original}} \times 100$$

Name _____

Wire Sample Data

Wire Diameter: 0.040″ (1.02 mm)

Area (in^2 or mm^2): _____

Percent Cold-Worked: _____

Drawn Specimen

Max Load (lbf or N): _____

UTS (lbf/in^2 or N/mm^2 (or Pa)): _____

Change in Length (in or mm): _____

Percent Elongation: _____

Annealed Specimen

Max Load (lbf or N): _____

UTS (lbf/in^2 or N/mm^2 (or Pa)): _____

Change in Length (in or mm): _____

Percent Elongation: _____

Wire Diameter: 0.035″ (0.89 mm)

Area (in^2 or mm^2): _____

Percent Cold-Worked: _____

Drawn Specimen

Max Load (lbf or N): _____

UTS (lbf/in^2 or N/mm^2 (or Pa)): _____

Change in Length (in or mm): _____

Percent Elongation: _____

Annealed Specimen

Max Load (lbf or N): _____

UTS (lbf/in^2 or N/mm^2 (or Pa)): _____

Change in Length (in or mm): _____

Percent Elongation: _____

Wire Diameter: 0.032" (0.81 mm)

Area (in² or mm²): _____

Percent Cold-Worked: _____

Drawn Specimen

Max Load (lbf or N): _____

UTS (lbf/in² or N/mm² (or Pa)): _____

Change in Length (in or mm): _____

Percent Elongation: _____

Annealed Specimen

Max Load (lbf or N): _____

UTS (lbf/in² or N/mm² (or Pa)): _____

Change in Length (in or mm): _____

Percent Elongation: _____

Wire Diameter: 0.028" (0.71 mm)

Area (in² or mm²): _____

Percent Cold-Worked: _____

Drawn Specimen

Max Load (lbf or N): _____

UTS (lbf/in² or N/mm² (or Pa)): _____

Change in Length (in or mm): _____

Percent Elongation: _____

Annealed Specimen

Max Load (lbf or N): _____

UTS (lbf/in² or N/mm² (or Pa)): _____

Change in Length (in or mm): _____

Percent Elongation: _____

Name _____

Wire Diameter: 0.025" (0.64 mm)

Area (in² or mm²): _____

Percent Cold-Worked: _____

Drawn Specimen

Max Load (lbf or N): _____

UTS (lbf/in² or N/mm² (or Pa)): _____

Change in Length (in or mm): _____

Percent Elongation: _____

Annealed Specimen

Max Load (lbf or N): _____

UTS (lbf/in² or N/mm² (or Pa)): _____

Change in Length (in or mm): _____

Percent Elongation: _____

Review Questions

1. Did the ultimate tensile strength of the cold-worked wire change as you reduced the diameter and increased the amount of cold work? If so, how did it change?

   _____
   _____
   _____
   _____
   _____
   _____

2. Did the ultimate tensile strength of the annealed wire change as you reduced the diameter? If so, how did it change?

   _____
   _____
   _____
   _____
   _____
   _____
   _____

3. You reduced the wire diameter by a series of draws, or reductions. What would happen if you tried to reduce the 0.040″ wire to 0.025″ in one draw?

___
___
___
___
___
___
___